STERLING
Education

Essential Chemistry

Stoichiometry

4th edition

4 3 2 1

ISBN-13: 979-8-8855705-9-6

Sterling Education materials are available at quantity discounts.

Contact info@sterling–prep.com

Sterling Education
6 Liberty Square #11
Boston, MA 02109

Published by Sterling Education

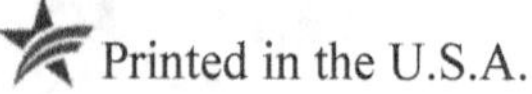 Printed in the U.S.A.

STERLING
Education

From the foundations of chemical reactions to the complex mechanisms of atomic particles, *Essential Chemistry Self-Teaching Guides* are a comprehensive compendium of clearly explained texts to learn and master these multifaceted chemistry topics.

These guides provide a detailed review of the fundamental mechanisms of chemical and physical processes at the atomic level. Develop a better understanding of the electronic structure of elements, principles of chemical bonding, phases of matter, types and mechanisms of chemical reactions, and principles of solutions and acid-base equilibria. Learn about rate processes in chemical reactions, empirical and molecular formulas, enthalpy, entropy, oxidation number, the laws of thermodynamics, and electrochemistry. Reinforce your learning by working through the practice questions and step-by-step solutions.

Created by highly qualified chemistry instructors, researchers, and education specialists, these books empower readers by helping them increase their understanding of general chemistry.

We sincerely hope that these guides are valuable for your learning.

250909akp

Featured on

Electronic Structure & Periodic Table

Chemical Bonding

States of Matter & Phase Equilibria

Stoichiometry

Solution Chemistry

Chemical Kinetics & Equilibrium

Acids & Bases

Chemical Thermodynamics

Electrochemistry

Visit our Amazon store

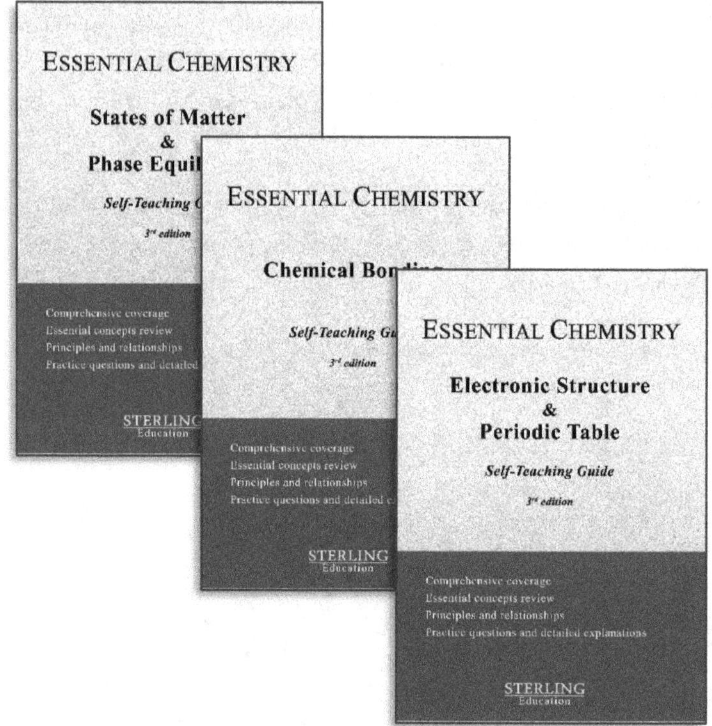

Essential Physics Self-Teaching Guides

Kinematics and Dynamics

Equilibrium and Momentum

Force, Motion and Gravitation

Work and Energy

Fluids and Solids

Waves and Periodic Motion

Light and Optics

Sound

Electrostatics and Electromagnetism

Electric Circuits

Thermal Physics

Atomic and Nuclear Physics

Visit our Amazon store

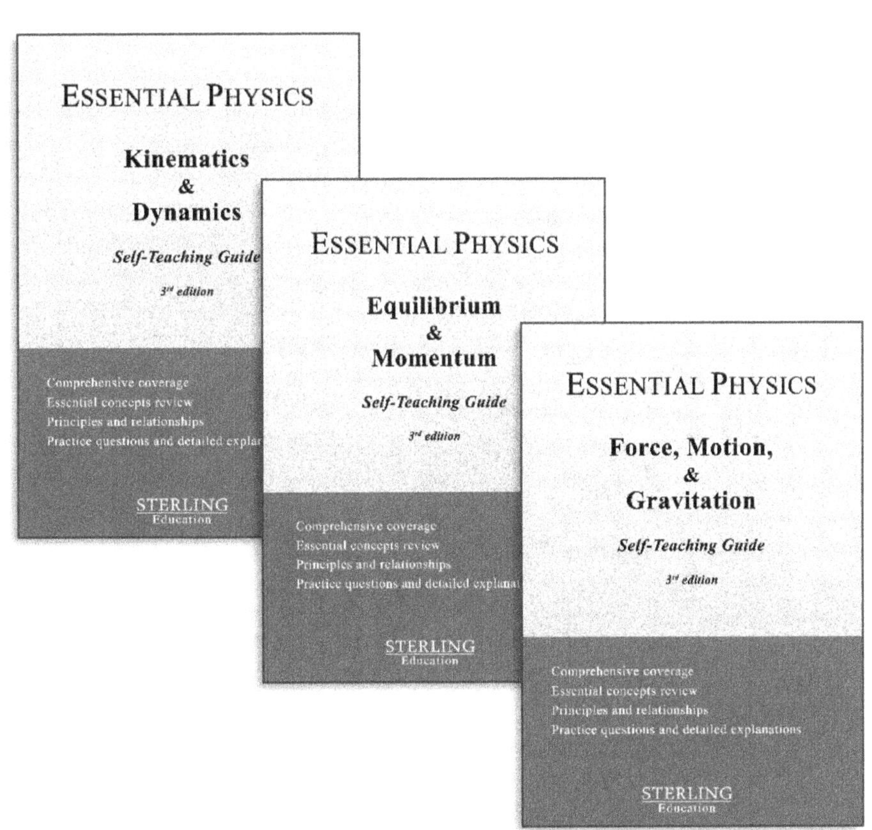

Essential Biology Self-Teaching Guides

Eukaryotic Cell & Cellular Metabolism

Molecular Biology & Genetics

Nervous & Endocrine Systems

Circulatory, Respiratory & Immune Systems

Digestive & Excretory Systems

Muscle, Skeletal & Integumentary Systems

Reproduction & Development

Microbiology

Plants & Photosynthesis

Evolution, Classification & Diversity

Ecology & Population Biology

Visit our Amazon store

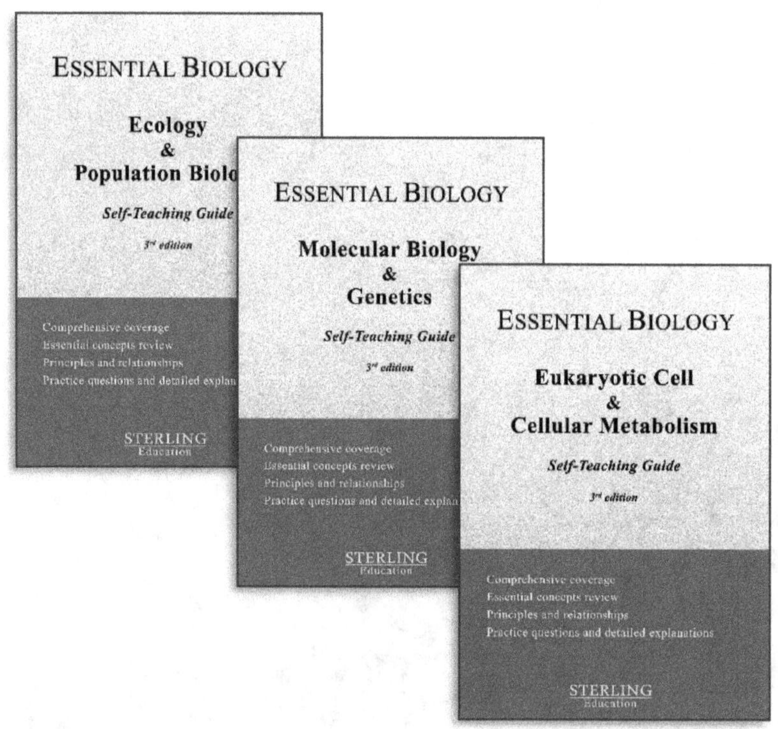

For online practice resources visit

https://www.sterling-prep.com

If you benefited from this book, please leave a review on Amazon so others
can learn from your input. Reviews help us understand our customers'
needs and experiences while keeping our commitment to quality.

Table of Contents

Table of Contents (*continued*)

Table of Contents (*continued*)

Page intentionally left blank

Page intentionally left blank

The Nature of Science

Scientific discovery

Observation is the first step toward scientific discovery. Observations recognize patterns and anomalies in unexplained phenomena.

Hypothesis is advanced as a proposed explanation for the observation.

Evaluating a hypothesis involves experiments and additional observations (i.e., recording and collecting data).

Experimental observations are made while manipulating *independent variables* (e.g., how long the plant is exposed to sunlight per twenty-four hours) with *dependent variables* (data collected regarding height measurements).

Observations generate *data* that:

    1) *support the hypothesis* (i.e., different from proves) or

    2) *refute the hypothesis*.

Hypothesis are modified by narrowing or expanding their application to explain additional data. More experiments are performed, and the results either *support* (not prove) or *refute* the hypothesis during this iterative process.

Principles are the initially proposed explanations that are specific and apply to a narrow range of phenomena.

Theories build upon valid hypotheses

Theories are proposed to account for the observations when a hypothesis is valid over a range (e.g., sunlight, temperature, or humidity).

Theories are advanced to explain the phenomena and to predict what will happen under other conditions with different variables.

Further observations determine whether the predictions were accurate.

Research (re-search or *again search*) continues as these many observations support hypotheses and theories.

When theories are accurate over various scenarios and experiments, *Laws* are developed based on these repeatedly replicated results.

Laws as established theories

Laws are brief descriptions of how nature behaves in a broad set of circumstances. They are the consolidation of several theories. Laws are the products of multiple theories tested and proven, culminating in a broad claim regarding those predictions.

Laws are widely applicable explanations. Laws with broad and straightforward claims are more robust than those that make claims on a narrower, more specific set of variables and parameters.

The number of Laws describing physical phenomena is minuscule compared to the number of theories and magnitudes, which are less than the number of supported hypotheses.

The number of refuted (unsupported) hypotheses may be several magnitudes larger.

Failed experiments do not typically refer to the physical process where the researcher made a human error during the experiment. A failed experiment means that the *data does not support the hypothesis.* Therefore, after additional trials to validate the "failed" results, the hypothesis must be modified to predict the results of either future experiments or objective observations in the physical world.

Theories, models and laws

Theories emerge from a hypothesis that experimental data has repeatedly supported. Theories are detailed statements that provide testable predictions of the behavior of natural phenomena.

Theories are established to explain observations and then tested (supported or refuted) based on their predictions. Tested theories can articulate the boundaries of the tested hypothesis.

Theories are more specific than *hypotheses* but vaguer and more encompassing than *models*.

Models are constructed to explain the observed behavior if a hypothesis accurately predicts a phenomenon. They are more specific in their application than theories.

Data collected to develop the theories provide the conceptual foundation for constructing models. The model creates mental pictures (e.g., complicated weather patterns displayed as visual pictures for the viewer) of the physical phenomena.

Models should be consistent with scientific theories' predictions and support a comprehensive understanding of a phenomenon. Models allow for predictive outcomes under specific theories (e.g., when the wind speed reaches x, then y will occur).

Multiple models are integrated to explain a portion, or the entire scope, of a *theory*.

Understand the limitations of a model and do not apply it dogmatically unless its applicability is evaluated. A model is an underlying representation of a part of a theory; it is not intended to provide a complete picture of every phenomenon under that theory.

Instead, models provide the fundamental aspects of the theory.

Chemical Formulae and Molecular Mass

Chemical formula

Chemical formulas are commonly expressed as molecular and empirical formulas.

Molecular formula describes the atomic composition of a molecule in its naturally occurring form, specifying the number of each type of atom present.

Empirical formula describes the *simplest integer ratio* of atoms.

For example,

> *Molecular formula* of hydrogen peroxide is H_2O_2.

> *Empirical formula* of hydrogen peroxide is HO.

Molecular structure, molecular formula, and empirical formula of glucose are provided below.

Molecular Structure	Molecular Formula	Empirical Formula
CHO | H——C——OH | H——C——OH | H——C——OH | H——C——OH | CH_2OH	$C_6H_{12}O_6$	CH_2O

Molecular compounds

Molecular compounds are electrically neutral particles consisting of two or more nonmetals covalently bonded.

Covalent bonds involve the sharing of electrons between two atoms during chemical bonding.

Groups of atoms within a molecule behave like single particles or discrete units.

Based on the composition and type of atoms, molecules have a specific molecular weight (MW).

For example, a water molecule consists of two hydrogen atoms bonded to one oxygen atom.

Particle diagrams

A particle diagram has elements and compounds represented as distinctive shapes (e.g., balls, shaded or unshaded).

Particle diagrams can represent elements and compounds, as well as their molecular composition, by the types of shapes and how they are connected.

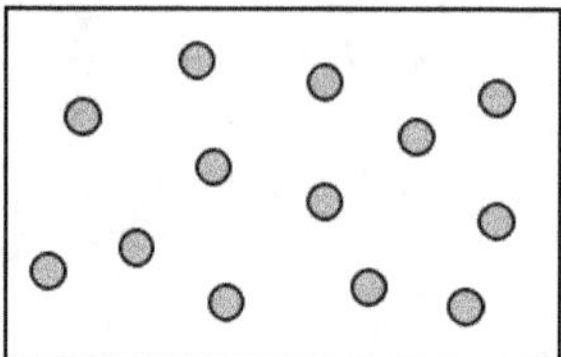 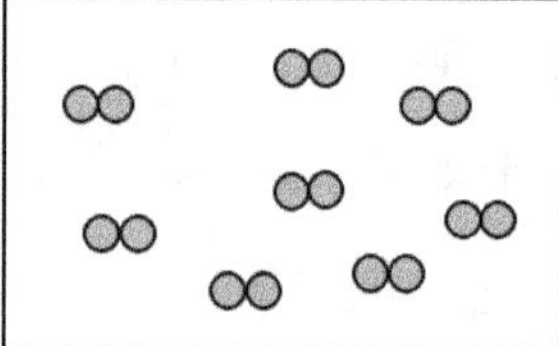

Elements

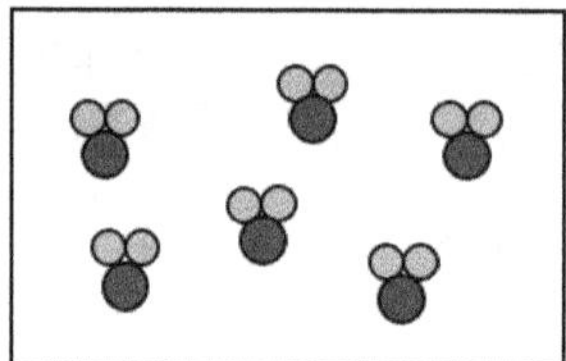 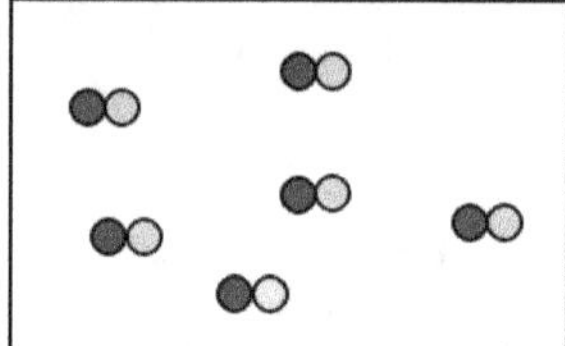

Compounds

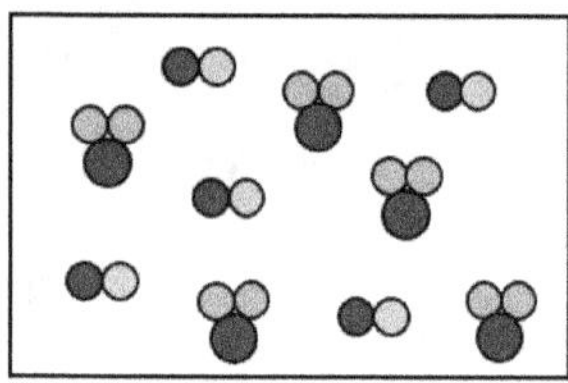 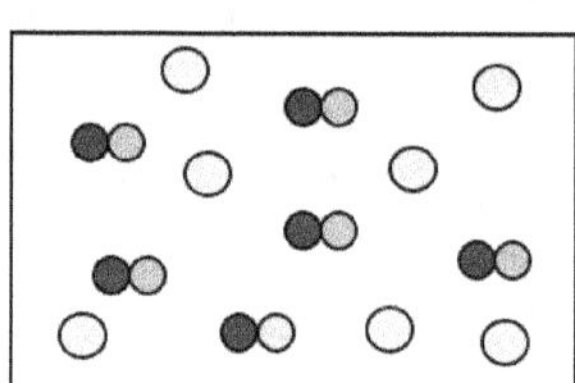 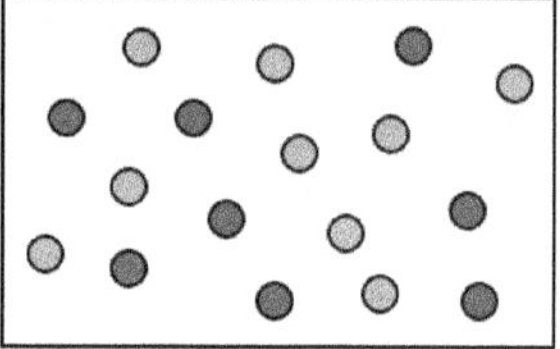

Mixtures

Avogadro's number

Molecular weight (MW) is the weight of 1 mole of molecules, where 1 mole equals 6.02×10^{23} particles.

The 6.02×10^{23} particles value is Avogadro's number (N_A).

Molecular weight of a substance is the sum of the atomic masses of its atoms. Typically, molecular weight is expressed in grams per mole (g/mol).

Molecular weight can be expressed in *atomic mass units* (amu or u), where 1 u = 1 g/mol.

> 1 amu = 1 dalton (Da).

For example, ^{12}C weighs 12 amu = 12 g/mol.

Atomic mass

Atomic mass is the mass of a single atom.

Molecular weight is the sum of the weights of the atoms in the molecule.

Atomic mass is listed on the periodic table; it is directly below the symbol for the element, as indicated by the arrow for lithium (Li).

Molar mass

Molar mass is the mass of a substance divided by the amount of a substance, expressed in g/mol.

Atomic mass refers to the mass of individual atoms, whereas molecular weight refers to the mass of a molecule.

Formula weight is the sum of the weights of the atoms in an empirical formula of a molecule.

For example, the molar mass of methane (molecular formula = CH_4)

$$CH_4 = 1 \text{ carbon } (12.0108 \text{ g/mol}) + 4 \text{ hydrogen } (1.0079 \text{ g/mol})$$

$$CH_4 = 12.0108 \text{ g/mol} + 4(1.0079 \text{ g/mol})$$

$$CH_4 = 16.0424 \text{ g/mol}$$

Molecular mass

Molecular mass (m) is the mass of a given molecule, measured in daltons (Da or u).

The molecular mass is the *ratio of the mass of a molecule to the atomic mass unit*. The molecular mass is more commonly used when referring to the mass of a single molecule.

The molecular mass is often used interchangeably with molecular weight (MW).

Molecules of the same compound have different molecular masses due to different isotopes of an element.

Molecular weight

Molecular weight (MW) is the sum of the weights of the atoms in a molecular formula.

Molecular weight commonly refers to a *weighted average* of a sample.

For example, determine the molecular weight of water (H_2O).

Water, as indicated by its chemical formula, has two hydrogen atoms and one oxygen atom.

From the periodic table, the mass of a hydrogen (H) atom is 1.008 amu, and that of oxygen (O) is 15.999 amu.

For example, the equation to determine the MW of water (H_2O).

Chemical formula for water:

2 (H) + 1 (O)

Molecular weight for water:

MW= 2 H (1.008 g/mol) + 1 O (15.999 g/mol)

MW = 18.015 g/mol

Another example of atomic mass is seen in the graphic below.

A single carbon atom is approximately 12 amu (12 g/mol).

Oxygen exists as a diatomic molecule.

O_2 (2 × 15.999 g/mol = 32 g/mol)

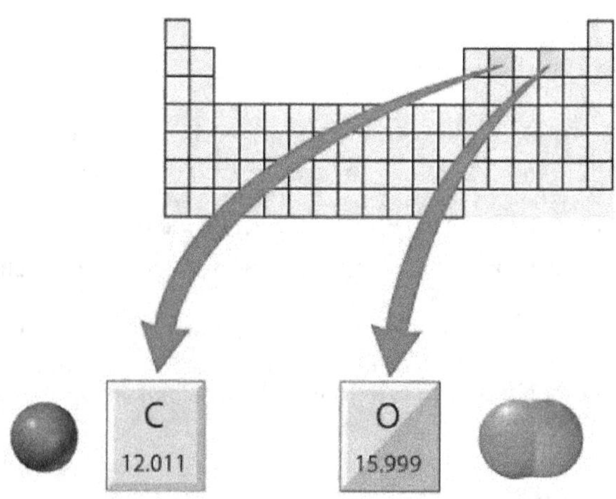

The mass of one carbon atom is approximately 12 amu.

The mass of an O_2 molecule is approximately (16 + 16) 32 amu.

A carbon (C) atom is [12 g/mol] / (32 g/mol)], or 3/8 as massive as an O_2 molecule.

Naming Molecular Compounds

Naming binary molecular compounds

Binary molecular compounds are covalently bonded with two elements, generally nonmetals.

Writing the molecular formula of a binary molecular compound follows three rules:

1. The first element in the formula is first, using the element's full name.

2. The second element is named as if it were an anion.

3. A prefix indicates the number of each element in a compound (see the table below).

Numerical prefixes for naming compounds

Prefix	Meaning
mono-	1
di-	2
tri-	3
tetra-	4
penta-	5
hexa-	6
hepta-	7
octa-	8

For example, the formula for carbon dioxide is CO_2.

The name carbon dioxide indicates a single carbon (no prefix because the *mono-* prefix is omitted when the first element exists as a single atom).

The name carbon dioxide indicates two oxygens due to the prefix *di- (meaning two)*.

Disulfur tetrafluoride has the formula S_2F_4:

di- prefix indicates two sulfur atoms.

tetra- prefix indicates four fluorine atoms.

Naming ionic compounds

Ionic compounds consist of a metal atom and a nonmetal atom that transfers valence electrons and is bonded by electrostatic forces between oppositely charged particles.

Ionic compounds are named differently. Identify the ions (e.g., cations, monatomic, polyatomic anions) for writing a chemical formula from an ionic compound. The inability to recognize these ions is a leading cause of difficulty in writing chemical formulae of inorganic compounds; memorization of the ions is essential.

The cation is listed first and the anion second in an ionic compound (or chemical formula).

For example, sodium chloride (NaCl) forms from a sodium cation (Na^+) and chlorine anion (Cl^-).

Some cations form one ion (Group 1 (IA), 2 (IIA), and 3 (IIIA) elements, except Thallium). Cations that form more than one ion have a stock number, denoted by a Roman numeral in parentheses (e.g., Cu(I), Fe(II), or Fe(III)).

For monatomic anions, the charge equals the Group number minus 8.

For example, oxygen has the monatomic anion O^{2-}.

Oxygen's group number is 6, so the charge is $6 - 8 = -2$.

Common polyatomic ions

$C_2H_3O_2^-$	acetate	OH^-	hydroxide
NH_4+	ammonium	ClO^-	hypochlorite
CO_3^{2-}	carbonate	NO_3^-	nitrate
ClO_3^-	chlorate	NO_2^-	nitrite
ClO_2^-	chlorite	$C_2O_4^{2-}$	oxalate
CrO_4^{2-}	chromate	ClO_4^-	perchlorate
CN^-	cyanide	MnO_4^-	permanganate
$Cr_2O_7^{2-}$	dichromate	PO_4^{3-}	phosphate
HCO_3^-	bicarbonate	SO_4^{2-}	sulfate
HSO_4^-	bisulfate	SO_3^{2-}	sulfite
HSO_3^-	bisulfite		

Learn the polyatomic anions in the table above

Common Metric Units

Seven base units of SI measurements

Metric system is a decimal system of measurement, known as the *International System of Units* (abbreviated as SI).

SI has seven base units:

> length (meters)
>
> mass (kilograms)
>
> time (seconds)
>
> electric current (amperes)
>
> temperature (Kelvin)
>
> amount of substance (moles)
>
> luminous intensity (candelas)

SI units of measurement, such as volume (liters), density (kg/m^3), and pressure (atmospheres), are derived from the base units.

SI units are based on powers of 10 (or multiples of 10); conversion between units is uniform and straightforward. The SI system uses prefixes to indicate the magnitude of a measured quantity; these prefixes provide the conversion factor.

Common prefixes, symbols, and powers

Learn the common prefixes shown:

Prefix	Symbol	Power	Prefix	Symbol	Power
mega-	M	10^6	centi-	c	10^{-2}
kilo-	k	10^3	milli-	m	10^{-3}
hecto-	h	10^2	micro-	μ	10^{-6}
deca-	D	10^1	nano-	n	10^{-9}
deci-	d	10^{-1}	pico-	p	10^{-12}

SI units do not allow for double prefixes (e.g., $1{,}000\ g \neq 1$ hectodecagram); instead, one prefix is used for the quantity of base units (e.g., $1{,}000\ g = 1$ kilogram).

Common conversions

Memorize a few typical metric-imperial conversions:

Length: 2.54 cm = 1 inch

Mass: 454 g = 1 pound

Volume: 0.946 L = 1 quart

Temperature: °C = (°F − 32) / 1.8

Significant figures

Significant figures indicate the number of digits known for a measured (or calculated) quantity within a *degree of uncertainty*.

For example, if a thermometer indicates a boiling point of 36.2 °C and has an uncertainty of ± 0.2 °C, three figures in 36.2 are significant (including the 0.2, which is uncertain).

The notation 36.2 ± 0.2 °C indicates the known quantity and the uncertainty digits (± 0.2 °C).

Converting between units

Conversions between metric units involve adding or subtracting zeros.

For example, suppose the mass of a 250 mg aspirin tablet needs to be converted to grams. Start by using the units to set up the problem. If a unit is to be converted (in this case, mg), it is placed in the numerator. That unit must be in the denominator of the conversion factor for it to cancel:

$$(250 \text{ mg}/1) \cdot (1 \times 10^{-3} \text{ g}/1 \text{ mg})$$

$$(250 \text{ mg}/1) \cdot (1 \times 10^{-3} \text{ g}/1 \text{ mg}) = 0.250 \text{ g}$$

Units cancel to give grams.

The conversion factor is shown as a numerator of 1×10^{-3} as it must be entered on most calculators, not 10^{-3}.

The mg is a value of 1, and the prefix "*milli*" applies to the gram unit. 1 mg = 1×10^{-3} g.

Conversions between English/imperial and metric units work similarly.

The difference lies in the conversion factors, which are not powers of ten and vary for each unit, making conversions between English/imperial and metric units more complex.

Therefore, the imperial system is not ideal for scientific calculations.

Converting with multiple units

For example, convert the mass of a 23-pound object to kilograms.

Pound units convert to grams, and grams are converted to kilograms.

Use the proper units to solve the problem:

$$23 \text{ lbs.} = (23 \text{ lbs.} / 1) \times (454 \text{ g}/ 1 \text{ lbs.}) \times (1 \text{ kg}/1 \times 10^3 \text{ g})$$

$$23 \text{ lbs.} = (23 \cancel{\text{ lbs.}} / 1) \times (454 \cancel{\text{ g}}/ 1 \cancel{\text{ lbs.}}) \times (1 \text{ kg}/1 \times 10^3 \cancel{\text{ g}})$$

$$23 \text{ lbs.} = 10 \text{ kg}$$

A conversion problem may include multiple units.

For example, convert the pressure of 14 lb/in² to g/cm².

For problems involving multiple conversions, work with one unit at a time.

Convert the pounds to gram units:

$$14 \text{ lb/in}^2 \times 454 \text{ g/lb}$$

Convert the in^2 unit to cm^2 units.

Set up the conversion without the exponent first, using the conversion factor:

$$1 \text{ in} = 2.54 \text{ cm}$$

Since in^2 and cm^2 are needed, raise each value to the second power:

$$= 14 \text{ lb/in}^2 \times 454 \text{ g/lb} \times 1^2 \text{ in}^2/2.54 \text{ cm}^2$$

$$= 9.9 \times 10^2 \text{ g/cm}^2$$

When units are squared, the numbers associated with them must be squared.

For 1 in^2, it does not make a difference because $1^2 = 1$.

Square the quantities (numbers) when the units are squared.

Verify that the units are set up correctly to ensure the problem has been configured properly.

Percent mass

Percent mass (% mass) is concentration *comparing the mass of one part of a substance to the mass of the whole.*

Percent mass unit is used for solutions, especially concentrated acid, or base solution, where the concentration is expressed as percent by mass on the bottle.

Percent mass is:

% mass = (mass of species of interest / total mass) × 100%

For example, find the % mass of sodium (Na) in sodium bicarbonate ($NaHCO_3$).

Use the periodic table to find the atomic mass of the elements.

Na = 22.99 g/mol

H = 1.01 g/mol

C = 12.01 g/mol

O = 16.00 g/mol

Use molecular mass of the atoms to calculate the molecular weight of $NaHCO_3$:

1 Na (22.99 g/mol) + 1 H (1.01 g/mol) + 1 C (12.01 g/mol) + 3 O (3 × 16.00 g/mol)

MW $NaHCO_3$ = 84.01 g/mol

Find percent mass of Na:

% mass Na = (mass of species of interest / total mass) × 100%

% mass Na = (22.99 g/mol / 84.01 g/mol) × 100%

% mass Na = 27.4%

Concentration can be expressed as a percentage volume (% volume), calculated similarly to percentage mass (% mass).

Mole Concept

Stoichiometry

Stoichiometry involves quantitative relationships between reactants and products.

The number of molecules is expressed in the mole unit.

Avogadro's number (N_A) is the number of particles in 1 mole of a substance and equals 6.02×10^{23} particles.

The substance's mass and moles are determined for reactants and products in a reaction.

$\underline{2\ H_2}$	+	$\underline{1\ O_2}$	$\rightarrow$	$\underline{2\ H_2O}$
2 moles = 4 g		1 mole = 32 g		2 moles = 36 g
12.04×10^{23} molecules		6.02×10^{23} molecules		12.04×10^{23} molecules

Most stoichiometry problems follow a set strategy using moles:

Quantity A $\rightarrow$ Moles A $\rightarrow$ Moles B $\rightarrow$ Quantity B

Many stoichiometry problems are solved using this strategy.

Each step is examined and combined to solve complicated problems.

Moles to moles conversion

For example, how many moles of $CaCO_3$ are in a 25.0 g sample?

Calculate molar mass of $CaCO_3$ using atomic mass information from the periodic table.

Calculate molar mass with the same significant figures as the quantity converted:

1 Ca = 40.08 g/mol $\times$ 1 = 40.08 g/mol

1 C = 12.01 g/mol $\times$ 1 = 12.01 g/mol

3 O = 16.00 g/mol $\times$ 3 = 48.00 g/mol

$CaCO_3$ = 40.08 g/mol + 12.01 g/mol + 48.00 g/mol

$CaCO_3$ = 100.09 = 100.1 g/mol

Use molar mass to convert the 25.0 g mass of $CaCO_3$ to moles $CaCO_3$.

$CaCO_3$ = 25.0 g $\times$ (1 mol $CaCO_3$ / 100.09 g $CaCO_3$)

$CaCO_3$ = 0.250 mol

Moles to grams conversion

For example, what is the number of grams of $CaCO_3$ in 0.750 mol $CaCO_3$? (Use molecular mass of $CaCO_3$ = 100.09)

$CaCO_3$ = 0.750 mol × (100.09 g $CaCO_3$ / 1 mol $CaCO_3$)

$CaCO_3$ = 75.1 g $CaCO_3$

Moles A to moles B conversion

For example, how many moles of sodium ions (Na^+) does 0.100 mol of sodium carbonate (Na_2CO_3) have?

One mole of sodium carbonate, Na2CO3, contains 2 moles of Na, 1 mole of C, and 3 moles of O. Na2CO3 completely dissociates (i.e., it is an electrolyte) into ions in a solution. Hence, the sodium atoms are present as ions. Compare the moles of Na_2CO_3 and the moles of Na^+:

= 0.100 mol Na_2CO_3 × (2 mol Na^+ / 1 mol Na_2CO_3)

= 0.200 mol Na^+

Moles of reactants to moles of products conversion

A reaction may involve other examples, and a comparison of two compounds is necessary.

The stoichiometric coefficient is the mole ratio between species for a balanced reaction.

For example, a typical dissociation reaction:

$$2\ KClO_3 \quad \rightarrow \quad 2\ KCl + 3\ O_2$$

In this reaction, 2 moles of potassium chlorate ($KClO_3$) decompose into 2 moles of potassium chloride (KCl) and 3 moles of oxygen (O_2).

How many O_2 molecules are produced during the dissociation reaction above?

If there are 0.400 mol of $KClO_3$ and the number of moles of O_2 is needed, use stoichiometric coefficients to arrange a mole ratio, so moles of $KClO_3$ cancel, and moles of O_2 remain:

= (0.400 mol $KClO_3$) × (3 mol O_2 / 2 mol $KClO_3$)

= 0.600 mol O_2

Avogadro's number converts the number of moles to the number of particles.

= 0.600 mol O_2 × (6.02 × 10^{23} molecules O_2 / 1 mol O_2)

= 3.61 × 10^{23} O_2 molecules

Mole units of O_2 successfully cancel in the answer.

Density and Concentration

Density and specific gravity

Density (ρ) is the ratio of mass to volume of a substance.

SI unit of density is kilograms per cubic meter (kg/m^3).

Specific gravity is a unitless ratio between the density of a sample and a *reference substance* (often water).

Some units of density:

$$\text{Density of water} = 1 \text{ g/mL} = 1 \text{ g/cm}^3$$

$$\text{Specific gravity of water} = 1 \text{ g/cm}^3 / 1 \text{ g/cm}^3 = 1$$

$$\text{Density of lead} = 11 \text{ g/cm}^3$$

$$\text{Specific gravity of lead} = 11 \text{ g/cm}^3 / 1 \text{ g/cm}^3 = 11$$

The densest known element is osmium at 22.59 g/cm^3.

The least dense known element is hydrogen at $8.99 \times 10^{-5} \text{ g/cm}^3$.

Osmium is approximately 250,000 times denser than hydrogen.

Molarity and molality

Molarity (M) is the *moles of solute per liter of solution.*

$$\text{Molarity} = \text{moles of solute} / \text{L of solution}$$

Molarity is the most common concentration unit used for solutions.

Molality (m) is the moles of solute per kilogram of solvent (Phase Equilibria chapter).

$$\text{Molality} = \text{moles of solute} / \text{kg of solvent}$$

Concentration of dilute solutions

Very dilute solutions use the unit of *parts per million* (ppm).

$$\text{ppm} = \frac{\text{grams of solute}}{\text{grams of solution}} \times 10^6$$

The amount of solute relative to the amount of solvent is typically minimal.

Therefore, the density of the solution is, to a first approximation, the same as the density of the solvent.

Parts per million may be expressed as:

$$\text{ppm} = \frac{\text{mg solute}}{\text{kg solution}}$$

If the solvent is water (density of 1.00 kg/L), the expression for ppm:

$$\text{ppm} = \frac{\text{mg solute}}{\text{L solution}}$$

Measurement unit to express the concentration of dilute solutions is parts per billion (ppb).

Parts per billion may be expressed as:

$$\text{ppb} = \frac{\text{grams of solute}}{\text{grams of solution}} \times 10^9$$

The density of a dilute solution approximates the solvent's density.

Thus, parts per billion may be expressed as:

$$\text{ppb} = \frac{\mu\text{g solute}}{\text{kg solution}}$$

$$\text{ppb} = \frac{\mu\text{g solute}}{\text{L solution}}$$

Mole fraction and mole percent

Mole fraction (X) is the ratio of the moles of a specific molecule to the total moles of components in the mixture.

Mole percent ($\%X$) is the mole ratio multiplied by 100.

$$X_{\text{solute}} = \frac{\text{mol solute}}{\text{total moles of all components}}$$

$$X_{\text{solute}}\ \% = \frac{\text{mol solute}}{\text{total moles of all components}} \times 100$$

When converting between units, start by choosing an arbitrary amount of solution in the denominator of the concentration to be converted.

Percent mass to molarity

For example, if converting percent mass to molarity, assume 100 grams of solution.

If converting molarity to percent mass, assume one liter of solution.

For example, what are the molarity, molality, and mole fraction of HCl for a concentrated HCl solution known to be 37.0% HCl by mass, and its density is 1.19 g/ml?

Begin with the valid assumption that the HCl solution is 100 g.

Since the HCl percent mass is 37%, 37.0 g of the solution is HCl (grams of solute), the remaining 63.0 g is water (grams of solvent).

To find molarity, determine the moles of HCl (solute) per liter of solution.

First, convert mass of HCl to moles:

$$\text{mol HCl} = 37.0 \text{ g HCl} \times 1 \text{ mol HCl} / 36.5 \text{ g HCl}$$

$$\text{mol HCl} = 1.01 \text{ mol HCl}$$

Using the solution's density, convert the known mass of the solution (100 g) to liters of solution.

$$\text{L solution} = 100 \text{ g solution} \times \frac{1 \text{ mL solution}}{1.19 \text{ g solution}} \times \frac{1 \text{ L solution}}{1000 \text{ mL solution}}$$

$$\text{L solution} = 100 \,\cancel{\text{g solution}} \times \frac{1 \,\cancel{\text{mL solution}}}{1.19 \,\cancel{\text{g solution}}} \times \frac{1 \text{ L solution}}{1000 \,\cancel{\text{mL solution}}}$$

$$\text{L solution} = 0.0840 \text{ L solution}$$

Using moles of solute (HCl) and volume of solution in liters, calculate the molarity (M) of the solution as moles of solute per liter of solution:

$$M = \frac{1.01 \text{ mol HCl}}{0.0849 \text{ L solution}}$$

$$M = 12.0 \ M \text{ HCl/L solution}$$

Molality to mole fraction

From above, find the molality of the HCl solution.

The moles of solute are known (1.01 mol HCl).

Determine the mass of the solvent (H_2O) in kilograms:

$$63.0 \text{ g } H_2O \times (1 \text{ kg } H_2O / 1{,}000 \text{ g } H_2O) = 0.0630 \text{ kg } H_2O$$

Using moles and mass of solvent, calculate molality (m or b):

$$m = 1.01 \text{ mol HCl} / 0.0630 \text{ kg } H_2O$$

$$m = 16.0 \text{ mol HCl/kg } H_2O$$

$$m = 16.0 \text{ m } H_2O$$

Finally, determine the mole fraction of HCl.

From before, there are 1.01 moles of HCl. Calculate moles of H_2O.

$$\text{mol } H_2O = 63.0 \text{ g } H_2O \times (1 \text{ mol } H_2O / 18.0 \text{ g } H_2O)$$

$$\text{mol } H_2O = 63.0 \text{ } \cancel{\text{g } H_2O} \times (1 \text{ mol } H_2O / 18.0 \text{ } \cancel{\text{g } H_2O})$$

$$\text{mol } H_2O = 63.0 \times (1 \text{ mol } H_2O / 18.0)$$

$$\text{mol } H_2O = 3.50 \text{ mol}$$

The moles of the molecules in the problem are known.

Calculate the mole fraction of HCl:

$$\chi_{solute} = \frac{\text{mol solute}}{\text{total moles of all components}}$$

$$\chi_{HCl} = 1.01 \text{ mol HCl} / [(1.01 \text{ mol HCl} + 3.50 \text{ mol } H_2O)]$$

$$\chi_{HCl} = 0.244$$

Oxidation States

Oxidation number

Oxidation number (or *oxidation state*) is the charge an atom has or appears to have when a set of rules counts the compound's electrons.

Oxidation numbers are typically used in ionic compounds for oxidation-reduction reactions, which involve the transfer of electrons.

For nonionic compounds, oxidation numbers determine if the compound's chemical formula is written correctly.

Assigning oxidation numbers

1. Oxidation state of an element is zero.

2. For main group metals (Groups 1-2, 13-18), the ion's charge equals the valence electrons or group number.

   Transition metals do not follow this rule and can exhibit multiple oxidation states.

   Oxidation states are sometimes not by group number

3. Oxygen ion in a compound is typically –2 except peroxide ion (O_2^{2-}), which is –1.

4. The sum of oxidation states equals the polyatomic ion's (or neutral molecule's) overall charge for polyatomic ions (or compounds).

Determining oxidation states

For example, determine the oxidation states of the elements in potassium permanganate ($KMnO_4$).

Solution: Apply rule two.

Since potassium is a Group IA alkali metal, its oxidation state is +1.

Assign x to Mn for now since manganese may exist in several oxidation states.

There are four oxygen atoms in the permanganate ion, with oxidation states of –2 per O atom.

Overall charge of the neutral compound equals zero:

$$K \quad Mn \quad O_4$$

$$+1 \quad x \quad 4(-2)$$

Algebraic expression is:

$$1 + x - 8 = 0$$

Solving for x gives the oxidation state of manganese:

$$x - 7 = 0$$

$$x = +7$$

$$K \quad Mn \quad O_4$$

$$+1 \quad +7 \quad 4(-2)$$

For example, what is the oxidation state of chromium in dichromate ion $Cr_2O_7^{2-}$?

Start by assigning an oxidation state of -2 to oxygen.

Since the oxidation state for chromium is not known, and two chromium atoms are present, assign the algebraic value of $2x$ for chromium:

$$Cr_2 \quad O_7^{2-}$$

$$2x \quad 7(-2)$$

Use the algebraic equation to solve for x.

Since the overall charge of the ion is -2, the expression is set equal to -2, rather than 0:

$$2x + 7(-2) = -2$$

Solve for x:

$$2x - 14 = -2$$

$$2x = 12$$

$$x = +6$$

Each chromium ion has an oxidation state of $+6$.

What are the oxidation states of the elements in the polyatomic compound $Fe_2(CO_3)_3$?

Two elements (iron and carbon) have more than one oxidation state.

When considering molecules formed from cations (positive ions) and anions (negative ions), the molecule is broken down into its constituent ions. Determine the charge on each ion.

For example, iron (Fe) has more than one oxidation state, but the carbonate ion has an oxidation state of -2 (CO_3^{2-}).

With this information, Fe's oxidation state can be determined:

$$Fe_2 \quad (CO_3)^3$$

$$2x \quad 3(-2)$$

$$2x - 6 = 0$$

$$2x = 6$$

$$x = 3$$

Each iron ion in the compound has an oxidation state of $+3$.

Next, consider the carbonate ion independent of the iron (III) ion:

$$CO_3^{2-}$$

$$x \quad 3(-2)$$

$$x - 6 = -2$$

$$x = +4$$

Solution: The oxidation state of carbon is $+4$, and each oxygen is -2.

Oxidation-reduction reactions

Oxidation-reduction reactions occur when electrons are transferred between two chemicals.

For an oxidation-reduction (or redox) reaction to proceed, one substance in a reaction is oxidized while another substance is reduced; reduction or oxidation processes cannot occur separately.

Oxidation is the loss of electrons, typically associated with metals, and may involve the addition of oxygen or the removal of hydrogen.

Reduction is the gain of electrons, typically observed in nonmetals, and may involve the addition of hydrogen or the removal of oxygen.

In biological systems, cells oxidize and reduce metals. Cytochrome c protein plays a vital role in ATP production. The cytochrome c protein contains a Fe^{2+} cation that undergoes oxidation to form Fe^{3+}, then is reduced to Fe^{2+}.

In organic reactions, look for the movement of oxygen and hydrogen.

Combustion (i.e., burning) is a reaction with oxygen as an example of a redox reaction.

Oxidizing and reducing agents

Oxidizing agent (or *oxidant*) is a substance that oxidizes another by removing electrons.

Oxidizing agent gains the removed electrons, thus *reducing itself.*

Thus, the oxidation number of the oxidizing agent becomes *less positive.*

Reducing agent is a substance that reduces another by donating its electrons to the substance being reduced.

Reducing agent *loses* these electrons and is *oxidized.*

Thus, the reducing agent's oxidation number becomes *more positive* (or less negative).

Identifying reduced and oxidized species

The number line below helps identify the reduced or oxidized substances and whether they are oxidizing or reducing agents, respectively.

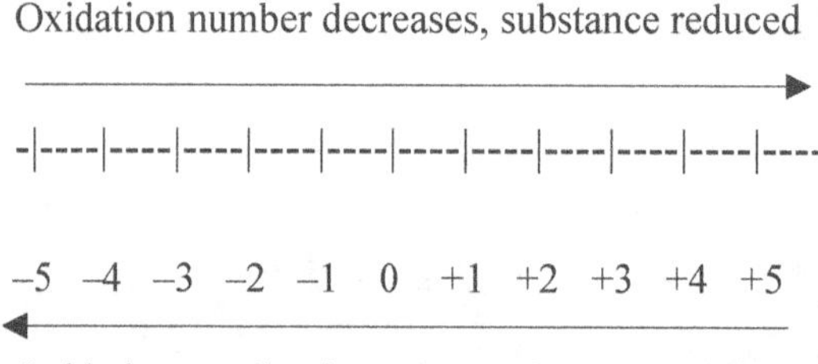

The image below summarizes the relationship between reducing agents and oxidizing agents.

Essentially, atoms that *lose electrons* during a chemical reaction undergo oxidation, and atoms that *gain electrons* during a chemical reaction undergo reduction.

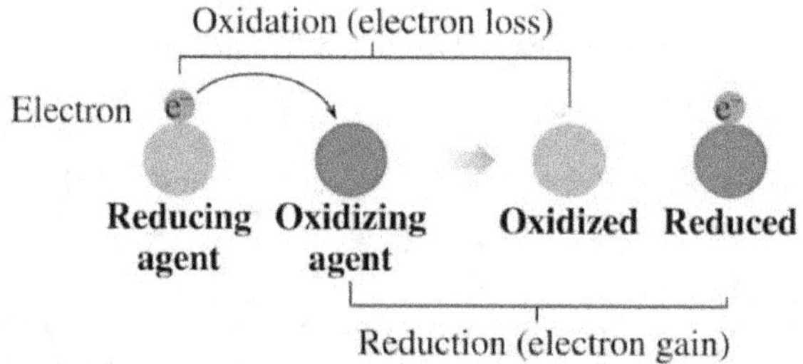

Oxidized *species* acts as the *reducing agent.*

Reduced species acts as the *oxidizing agent.*

Loss and gain of electrons

The OIL RIG mnemonic helps remember what is happening to the electrons in a redox reaction.

OIL RIG = **O**xidation **I**s **L**oss; **R**eduction **I**s **G**ain

Remember, this mnemonic device refers to the *loss and gain of electrons*.

For example. Identify the substance being oxidized, the substance being reduced, the oxidizing agent, and the reducing agent in the following reaction:

$$Cu \ (s) + 4 \ HNO_3 \ (aq) \rightarrow Cu(NO_3)_2 \ (aq) + 2 \ NO_2 \ (g) + 2 \ H_2O \ (l)$$

Determine the oxidation state for the reactants and products.

Elemental copper, Cu (s), has an oxidation state of zero (0).

For nitric acid (HNO_3) set N equal to x:

$$H \quad N \quad O_3$$

$$+1 \quad x \quad 3(-2)$$

$$1 + x - 6 = 0$$

$$x = +5$$

So

$$H = +1, \ N = +5, \ O = -2$$

For copper (II) nitrate, $Cu(NO_3)_2$:

$$Cu(NO_3)_2$$

Nitrates are -1, while Cu might be $+1$ or $+2$.

$$Cu(NO_3)_2$$

$$Cu \quad (NO_3)_2$$

$$x + 2(-1) = 0$$

$$x = 2$$

Cu's oxidation state is $+2$, oxygen is -2

For nitrate ion (NO_3^-) set N equal to x:

$$N \quad O_3^-$$

$$x \quad 3(-2)$$

$$x - 6 = -1$$

$$x = +5$$

So $Cu = +2$, $N = +5$, $O = -2$

For nitrogen dioxide (NO_2) set N equal to x:

N O_2

x $2(-2)$

$x - 4 = 0$

$x = +4$

So $N = +4$, $O = -2$

For water (H_2O):

H_2 O

$2(+1)$ -2

So $H = +1$, $O = -2$

The reaction is summarized in the following table.

$$Cu\ (s) + 4\ HNO_3\ (aq) \rightarrow Cu(NO_3)_2\ (aq) + 2\ NO_2\ (g) + 2\ H_2O\ (l)$$

Element	Oxidation State: reactants	Oxidation State: products
Cu	0	+2
H	+1	+1
N	+5	+5 (in NO_3^-), +4 (in NO_2)
O	-2	-2

Oxidation states of reactants and products for the example reaction above

Identify the molecules that changed their oxidation state.

Copper's oxidation number increased from 0 to +2, so copper has been oxidized (loses electrons), and therefore is the reducing agent.

Copper reduced nitric acid by donating electrons to the nitrogen in nitric acid.

Solution: Nitrogen in nitric acid changes from +5 to +4 in nitrogen dioxide, indicating that copper has reduced nitric acid. Since nitric acid is reduced, it is the oxidizing agent. Oxidation cannot occur without a reduction.

Common oxidizing and reducing agents

Common oxidizing agents	Common reducing agents
Oxygen O_2	Hydrogen H_2
Ozone O_3	metals (such as K)
Permanganates MnO_4^-	Zn/HCl
Chromates $CrO4_2^-$	Sn/HCl
Dichromates $Cr_2O_7^{2-}$	LAH (lithium aluminum hydride)
peroxides H_2O_2	$NaBH_4$ (sodium borohydride)
Lewis acids	Lewis bases
compounds with many oxygen atoms	compounds with many hydrogens

Common oxidizing and reducing agents

Disproportionation reactions

Disproportionation reactions involve an atom undergoing oxidation and reduction to form two atoms with different oxidation states.

For example, consider:

$$2\ Cu^+ \rightarrow Cu + Cu^{2+}$$

Cu^+ acts as both an oxidizing and reducing agent, simultaneously reducing and oxidizing itself.

Oxidized Cu^+ becomes Cu^{2+}

Reduced Cu^+ becomes Cu

For example, which of the following equations is a disproportionation reaction?

$$2\ H_2O \rightarrow 2\ H_2 + O_2$$

$$H_2SO_3 \rightarrow H_2O + SO_2$$

$$HNO_2 \rightarrow NO + HNO_3$$

$$Mg + H_2SO_4 \rightarrow MgSO_4 + H_2$$

Disproportionation is a redox reaction in which a species is simultaneously reduced and oxidized to form two distinct products.

Consider: $HNO_2 \rightarrow NO + HNO_3$

Unbalanced reaction: $HNO_2 \rightarrow NO + HNO_3$

Balanced reaction: $3\ HNO_2 \rightarrow 2\ NO + HNO_3 + H_2O$

The other reactions are classified as:

$2\ H_2O \rightarrow 2\ H_2 + O_2$: decomposition

$H_2SO_3 \rightarrow H_2O + SO_2$: decomposition

$Mg + H_2SO_4 \rightarrow MgSO_4 + H_2$: single replacement

Redox titration

Titration (or *volumetric analysis*) is a standard laboratory procedure used to determine the concentration of an unknown solu*tion*.

Titrant is the analytical reagent of known concentration carefully added to the sample from a buret until the *equivalence point* (i.e., the point of neutralization) is reached.

If the titrant's volume and concentration are known, its molar concentration can be calculated.

Endpoint is marked by a change in the indicator's color when a solution transitions from acidic to slightly basic.

Indicator is added to the reaction to determine the endpoint of a redox titration (i.e., when the sample molecules have been completely oxidized or reduced).

Redox indicator undergoes a color change at the neutralization reaction's equivalence point.

For example, ascorbic acid (vitamin C) titrations use iodine, which undergoes a redox reaction.

Ascorbic acid is *oxidized* (i.e., loses electrons) to dehydroascorbic acid, while iodine is *reduced* (i.e., gains electrons) to an iodide ion.

Starch is an *indicator* because excess iodine reacts with it, causing the solution to turn from clear to dark blue.

Ascorbic acid concentration is calculated using the initial volume of ascorbic acid (i.e., *analyte*) and iodine's measured volume (i.e., *titrant*).

Chemical Equations

Classifying chemical reactions

Predicting the products formed in each reaction is no easy task.

Since many chemical compounds exist, memorizing every reaction is not necessary.

Most chemical reactions are classified into several groups.

Identifying the reaction type is a crucial step in predicting the products of a chemical reaction.

The important classes of reactions encountered in general chemistry are shown:

Reaction Type	General Reaction
Synthesis	$A + B \rightarrow AB$
Decomposition	$AB \rightarrow A + B$
Replacement	$AB + C \rightarrow AC + B$ (single) $AB + CD \rightarrow AD + CB$ (double)

1. ***Combination reactions*** (or *synthesis reactions*): two (or more) substances form a single product.

   General form is $A + B \rightarrow AB$, in which two substances combine to form one compound.

2. ***Decomposition reactions***: one reactant forms two or more products, the reverse of a combination reaction.

   General form is $AB \rightarrow A + B$; a compound decomposes to form its constituent elements.

3. ***Replacement reactions*** (or *substitution reactions* and *displacement reactions*): one atom (or group) replaces another species in a compound.

   o In a ***single replacement reaction***, an element replaces another element of the same group.

   General form is $AB + C \rightarrow AC + B$, which replaces element B.

- In a ***double replacement reaction*** (also known as a *metathesis reaction*), two compounds react and exchange partners; specifically, cations and anions in one compound exchange with their counterparts in the other compound.

  General form is AB + CD $\rightarrow$ AD + CB

Note: In each type of replacement reaction, the number of substances on the reactant side of the equation is the *same* as on the product side.

Specific reaction types

- **Oxidation-reduction reactions** (or *redox*) *transfer electrons* between the two reactants.

  Loss of electrons (or an increase in oxidation number) is oxidation.

  Gain of electrons (or a decrease in oxidation number) is reduction.

  - Many combination and decomposition reactions utilize oxidation-reduction and single-displacement reactions.

- *Combustion* involves a carbon-containing molecule (e.g., hydrocarbon) reacting with oxygen to produce carbon dioxide (CO_2) and water (H_2O).

  - Combustion reactions tend to be *highly exothermic* and are considered *irreversible*.

  - Combustion is a decomposition and oxidation-reduction reaction.

  - Combustion utilizes oxygen (O_2) to produce energy, dissipating it as heat or performing work.

  - General formula for combustion of hydrocarbon:

  $$C_xH_y + O_2 \rightarrow CO_2 + H_2O$$

- *Condensation* produces water from a *double replacement reaction*.

  Condensation reactions occur among functional groups that contain ~H and ~OH that break from their compounds and form H_2O.

Adenosine triphosphate (ATP)

Adenosine diphosphate (ADP) + Energy

Hydrolysis (addition of water) for ATP conversion to ADP + energy (inorganic phosphate)

- *Hydrolysis* (hydro- means "*water*," lysis means "*break*") is the reverse process of condensation, where water is a reactant, while the other reactant molecule is split into two smaller product molecules.

The general form for *condensation and hydrolysis* is shown below.

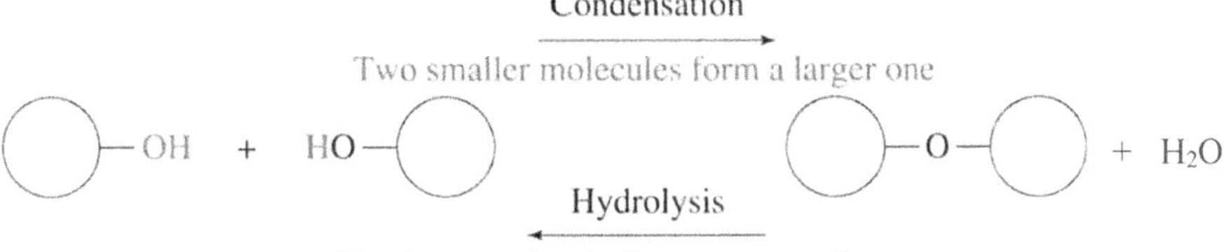

- *Carboxylation reactions* add a carboxyl (C=O) or carboxylic acid (~COOH) group.

   For example, as carbon dioxide (CO_2) moves through the cell, it is added to biomolecules by carboxylase and removed by decarboxylase (~*ase* for enzymes).

Carboxylation is shown below.

$$\text{bicarbonate} + \text{pyruvate} \xrightarrow[\text{carboxylase}]{\text{pyruvate}} \text{oxaloacetate} + H_2O$$

bicarbonate   pyruvate   oxaloacetate   water

Conventions for writing chemical equations

Chemical equations have several important parts:

$$H_2SO_4\ (aq) + 2\ NaOH\ (aq) \rightarrow 2\ Na^+\ (aq) + SO_4^{2-}\ (aq) + 2\ H_2O$$

phase   coefficient   direction   charge

Phase is indicated with a subscript, solid (s), liquid (l), gas (g), or aqueous (aq).

Coefficient indicates the relative number of moles of reactants or products.

Direction is represented as a single-headed arrow, denoting the forward direction.

Reversible reactions are at a state of *chemical equilibrium* and are represented with a double-headed arrow.

At equilibrium, the *rates* of the forward and reverse reactions are equal and constant, and no change in the *net number* of reactants or products.

A double-sided arrow with one side *larger* denotes a *nonequilibrium condition*, which spontaneously favors the larger arrow's direction.

Charge is indicated with a numerical superscript and +/– sign.

It is common not to indicate the charge on a neutral compound or substance.

Balancing chemical equations

Law of mass conservation (refer to the chapter on thermodynamics) states that atoms are neither created nor destroyed in a chemical reaction—they are simply rearranged.

The coefficients of the reactants and products must balance; there must be *equivalent amounts of atoms* on each side of the reaction.

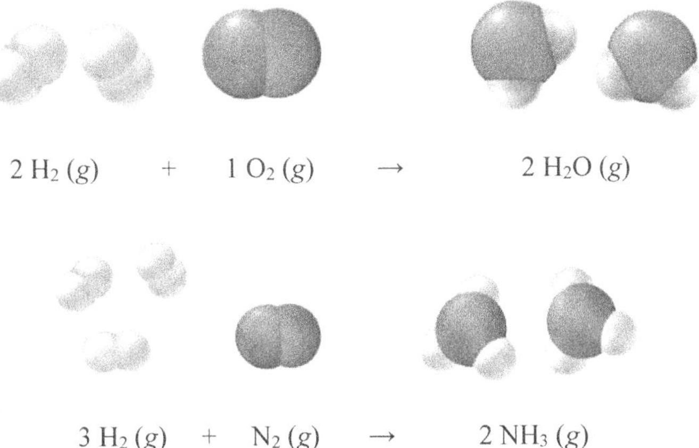

$$2\ H_2\ (g)\quad +\quad 1\ O_2\ (g)\quad \rightarrow \quad 2\ H_2O\ (g)$$

$$3\ H_2\ (g)\quad +\quad N_2\ (g)\quad \rightarrow \quad 2\ NH_3\ (g)$$

Balance a chemical equation by assigning coefficients to each species until the number of atoms is balanced on both sides.

For example, balance the combustion of propanol (C_3H_8O):

$$C_3H_8O + O_2 \rightarrow CO_2 + H_2O$$

Select the atom (or ion) present in one species on each side of the equation.

For the combustion of propanol, start with carbon.

Add the coefficient 3 to CO_2 on the right side since there are three carbons on the left side, indicated by C_3.

$$C_3H_8O + O_2 \rightarrow \mathbf{3}\ CO_2 + H_2O$$

Hydrogen is the other species in one molecule on each side, so it balances hydrogen.

Add the coefficient 4 to H_2O on the right because $4 \times 2 = 8$, which equals the number of H on the left, indicated by H_8.

$$C_3H_8O + O_2 \rightarrow 3\ CO_2 + \mathbf{4}\ H_2O$$

Starting with hydrogen before balancing carbon yields the same result.

First, balance elements present in one species on each side to avoid changing the coefficients of previously balanced species.

Oxygen is present every term, so it would be more difficult if O were balanced first.

Because of the adjustments to balance other atoms, one oxygen-containing species that has not yet been balanced.

Count the oxygen atoms:

one from C_3H_8O

six from $3\ CO_2$ (because $3 \times 2 = 6$)

four from $4\ H_2O$

Set equation:

$1 + 2x = (3 \times 2) + 4$, where x is the coefficient of the last term, O_2.

Solve for x, which equals $^9/_2$.

$C_3H_8O + x\ \mathbf{O_2} \rightarrow 3\ CO_2 + 4\ H_2O$

$C_3H_8O + ^9/_2\ \mathbf{O_2} \rightarrow 3\ CO_2 + 4\ H_2O$

Remove fractions, so multiply every term by 2.

$\mathbf{2}\ C_3H_8O + \mathbf{9}\ O_2 \rightarrow \mathbf{6}\ CO_2 + \mathbf{8}\ H_2O$

Remove non-integer coefficients, and the equation is balanced.

Redox Equations

Balancing redox reactions

When solving oxidation-reduction equations, a different approach is required.

Balancing redox reactions is more complicated and involves splitting the redox reaction into half-reactions.

Two methods for balancing half-reactions:

> *Ion-electron method* balances the elements by adding electrons to the charge.

> *Oxidation-state method* treats the species of interest as a single element (i.e., it changes oxidation number) and balances it accordingly.

Half-reactions balance oxidation-reduction reactions

Use the following procedure when balancing oxidation-reduction reactions:

1. Separate the equation into two *half-reactions*. A half-reaction contains only the species of interest (i.e., those containing the atom that changes oxidation state).

   Each half-reaction corresponds to the oxidation half-reaction or reduction half-reaction.

   Covalently attached moieties are not part of the species of interest.

   A species that does not change its oxidation state is a *spectator ion* (i.e., an ion present in solution but not involved in the reaction of interest).

2. Balance each half-reaction for the *charge* and *number* of atoms.

   Acidic conditions, add H_2O to the side that needs the oxygen atom, then add H^+ to the other side.

   Basic conditions: add $2\ OH^-$ to the side needing oxygen atoms, then add H_2O to the other.

3. After the half-reactions are balanced, recombine the half-reactions:

   Multiply each half-reaction by a factor so that the electrons cancel.

   Like solving a simultaneous equation, electron terms must be eliminated.

4. Lastly, perform these additional steps:

Combine identical species on the *same side* of the equation.

Cancel identical species on *opposite sides* of the equation.

Add back in the *spectator ions*.

For the oxidation-state method, balance oxygen and hydrogen elements.

Check that each side of the equation has an equal number of atoms and a neutral net charge.

For example, how many electrons are needed to balance the following half-reaction in a basic solution?

$$C_8H_{10} \rightarrow C_8H_4O_4{}^{2-}$$

Rules for balancing a half-reaction in basic conditions:

The first few steps are identical to those involved in balancing reactions in acidic conditions.

Step 1: Balance atoms except for H and O

$$C_8H_{10} \rightarrow C_8H_4O_4{}^{2-} \qquad \text{C is balanced}$$

Step 2: To balance oxygen, add H_2O to the side with fewer oxygen atoms

$$C_8H_{10} + 4\ H_2O \rightarrow C_8H_4O_4{}^{2-}$$

Step 3: To balance hydrogen, add H^+ to the opposing side of H_2O added in the previous step

$$C_8H_{10} + 4\ H_2O \rightarrow C_8H_4O_4{}^{2-} + 14\ H^+$$

This next step is the unique additional step for basic conditions.

Step 4: Add equal amounts of OH^- on both sides. The number of OH^- should match the number of H^+ ions. Combine H^+ and OH^- on the same side to form H_2O. If there are H_2O molecules on each side, subtract accordingly to end up with H_2O on one side only.

There are 14 H^+ ions on the right, so add 14 OH^- ions on both sides:

$$C_8H_{10} + 4\ H_2O + 14\ OH^- \rightarrow C_8H_4O_4{}^{2-} + 14\ H^+ + 14\ OH^-$$

Combine H^+ and OH^- ions to form H_2O:

$$C_8H_{10} + 4\ H_2O + 14\ OH^- \rightarrow C_8H_4O_4{}^{2-} + 14\ H_2O$$

H_2O molecules are on both sides, which cancel, and some H_2O remains on one side:

$$C_8H_{10} + 14\ OH^- \rightarrow C_8H_4O_4{}^{2-} + 10\ H_2O$$

Step 5: Balance charges by adding electrons to the side with a greater/more positive total charge

Total charge on left side: $14(-1) = -14$

Total charge on the right side: -2

Add 12 electrons to the right side:

$$C_8H_{10} + 14\ OH^- \rightarrow C_8H_4O_4^{2-} + 10\ H_2O + 12\ e^-$$

For example, how many electrons are needed to balance the charge for the following half-reaction in an acidic solution?

$$C_2H_6O \rightarrow HC_2H_3O_2$$

Balancing half-reactions in acidic conditions:

Step 1: Balance atoms except for H and O

$$C_2H_6O \rightarrow HC_2H_3O_2 \text{ (C is balanced)}$$

Step 2: To balance oxygen, add H_2O to the side with fewer oxygen atoms

$$C_2H_6O + H_2O \rightarrow HC_2H_3O_2$$

Step 3: To balance hydrogen, add H^+ to the opposing side of H_2O added in the previous step

$$C_2H_6O + H_2O \rightarrow HC_2H_3O_2 + 4\ H^+$$

Step 4: Balance charges by adding electrons to the side with a greater/more positive total charge

Total charge on the left side: 0

Total charge on right side: $4(+1) = +4$

Add 4 electrons to the right side:

$$C_2H_6O + H_2O \rightarrow HC_2H_3O_2 + 4\ H^+ + 4\ e^-$$

The reaction requires adding 4 electrons to the right side.

Ion-electron method for balancing redox reactions

For example, balance the following redox reaction using the ion-electron method:

$$K_2Cr_2O_7\ (aq) + HCl\ (aq) \rightarrow KCl\ (aq) + CrCl_3\ (aq) + H_2O\ (l) + Cl_2\ (g)$$

Step 1 – Separate into half-reactions:

$$\text{Reduction: } Cr_2O_7^{2-} \rightarrow Cr^{3+}$$

$$\text{Oxidation: } Cl^- \rightarrow Cl_2$$

The species of interest for the oxidation reaction is Cl^- (not HCl) because the H^+ is not covalently attached to Cl^- and the ions separate in an aqueous solution.

$Cr_2O_7^{2-}$ is used, not $K_2Cr_2O_7$, because K^+ and H^+ are spectator ions.

Step 2 – Balance each half-reaction:

For the ion-electron method, balance the elements first, then balance the charges.

Balance elements for the reduction half-reaction (ion-electron method):

$$Cr_2O_7^{2-} \rightarrow Cr^{3+}$$

$$Cr_2O_7^{2-} \rightarrow 2Cr^{3+}$$

$$Cr_2O_7^{2-} + 14\,H^+ \rightarrow 2\,Cr^{3+} + 7\,H_2O$$

Balance charge for the reduction half-reaction (ion-electron method):

$$Cr_2O_7^{2-} + 14\,H^+ + 6\,e^- \rightarrow 2\,Cr^{3+} + 7\,H_2O$$

Balance charge for the oxidation half-reaction (ion-electron method):

$$Cl^- \rightarrow Cl_2$$

$$2\,Cl^- \rightarrow Cl_2$$

$$2\,Cl^- \rightarrow Cl_2 + 2\,e^-$$

Step 3 – Recombine the half-reactions:

$$Cr_2O_7^{2-} + 14\,H^+ + 6\,e^- \rightarrow 2\,Cr^{3+} + 7\,H_2O$$

$$2\,Cl^- \rightarrow Cl_2 + 2\,e^-$$

Multiply each species in the second equation by 3:

$$6\,Cl^- \rightarrow 3\,Cl_2 + 6\,e^-$$

Add two equations:

$$Cr_2O_7^{2-} + 14\,H^+ + 6\,e^- + 6\,Cl^- \rightarrow 2\,Cr_3+ + 7\,H_2O + 3\,Cl_2 + 6\,e^-$$

Step 4 – Complete the process:

Except for electrons, there are no identical species to combine or cancel.

$$Cr_2O_7^{2-} + 14\,H^+ + 6\,Cl^- \rightarrow 2\,Cr^{3+} + 7\,H_2O + 3\,Cl_2$$

Step 5 – For the ion-electron method, the equation is balanced. The spectator ions need to be added to the equation.

When adding spectator ions, add equal numbers of ions to both sides of the reaction.

To the left side: the dichromate ion was paired with K^+, so add 2 K^+ for the dichromate.

To the right side: match the left side by adding 2 K^+.

$$K_2Cr_2O_7 + 14\ H^+ + 6\ Cl^- \rightarrow 2\ Cr^{3+} + 7\ H_2O + 3\ Cl_2 + \textbf{2 K}^+$$

Step 6 – There are 14 H^+ on the left side and 14 on the right, so they are balanced. Referring to the original equation, the H and Cl elements on the left originated from the HCl.

Add 8 Cl elements to the product side.

$$K_2Cr_2O_7 + 14\ HCl \rightarrow 2\ Cr^{3+} + 7\ H_2O + 3\ Cl_2 + 2\ K^+ + \textbf{8 Cl}^-$$

Step 7 – Right side shows two Cl^- ions to be combined with the 2 K^+, and the remaining 6 Cl^- ions go with the Cr.

The final balanced redox equation is:

$$K_2Cr_2O_7\ (aq) + 14\ HCl\ (aq) \rightarrow \textbf{2 CrCl}_3\ (aq) + 7\ H_2O\ (l) + 3\ Cl_2\ (g) + \textbf{2 KCl}\ (aq)$$

Oxidation-state method for balancing redox reactions

For example, balance the redox reaction (from above) using the oxidation-state method:

$$K_2Cr_2O_7\ (aq) + HCl\ (aq) \rightarrow KCl\ (aq) + CrCl_3\ (aq) + H_2O\ (l) + Cl_2\ (g)$$

Step 1 - Separate into half-reactions (same as the ion-electron method):

Reduction: $Cr_2O_7^{2-} \rightarrow Cr^{3+}$

Oxidation: $Cl^- \rightarrow Cl_2$

Step 2 - Balance each half-reaction:

Balance the elements of interest first when using the oxidation-state method.

Balance the elements for the reduction half-reaction (oxidation-state method):

1. $Cr_2O_7^{2-} \rightarrow Cr^{3+}$

2. $Cr_2O_7^{2-} \rightarrow 2\ Cr^{3+}$

3. Each oxygen is 2^- so the 2 Cr on the left must be 6^+

4. $2\ Cr^{6+} \rightarrow 2\ Cr^{3+}$

Balance charge for the reduction half-reaction (oxidation-state method):

1. $2\ Cr^{6+} + 6\ e^- \rightarrow 2\ Cr^{3+}$

Balance charge for the oxidation half-reaction (oxidation-state method):

1. $Cl^- \rightarrow Cl_2$

2. $2\ Cl^- \rightarrow Cl_2$

3. $2\ Cl^- \rightarrow 2\ Cl^\circ$

4. $2\ Cl^- \rightarrow 2\ Cl^\circ + 2\ e^-$

Step 3 - Recombine the half-reactions:

$$2\ Cr^{6+} + 6\ e^- \rightarrow 2\ Cr^{3+}$$

$$2\ Cl^- \rightarrow 2\ Cl^\circ + 2\ e^-$$

To cancel the electrons, multiply each term in the second equation by 3:

$$2\ Cr^{6+} + 6\ e^- \rightarrow 2\ Cr^{3+}$$

$$6\ Cl^- \rightarrow 6\ Cl^\circ + 6\ e^-$$

Add the two equations:

$$2\ Cr^{6+} + 6\ e^- + 6\ Cl^- \rightarrow 2\ Cr^{3+} + 6\ Cl^\circ + 6\ e^-$$

Step 4 - Except for the electrons, no like terms combine or cancel.

$$2\ Cr^{6+} + 6\ Cl^- \rightarrow 2\ Cr^{3+} + 6\ Cl^\circ$$

Convert the elements into species by referring to the original equation.

$$K_2Cr_2O_7 + 6\ HCl \rightarrow 2\ CrCl_3 + 3\ Cl_2$$

Unlike the ion-electron method, where the equation is balanced and spectator ions are added, the oxidation-state method requires balancing the equation again. After the elements are combined to recreate the molecules, the equation is no longer balanced.

Step 5 - Oxygen: There are seven O atoms on the left, so add 7 H_2O molecules to the right.

(Remember that this method applies to acidic reactions – see explanation for basic reactions.)

$$K_2Cr_2O_7 + 6\ HCl \rightarrow 2\ CrCl_3 + 3\ Cl_2 + \textbf{7 H}_2\textbf{O}$$

Step 6 - Hydrogen: There are 6 H atoms on the left, but 14 H atoms on the right.

Eight H atoms should be added to the left for 14 H atoms.

14 H atoms on the left should be HCl (refer to the original equation).

$$K_2Cr_2O_7 + \mathbf{14\ HCl} \rightarrow 2\ CrCl_3 + 3\ Cl_2 + 7\ H_2O$$

Important: HCl is the species of interest *and* spectator ion.

Some HCl contributes to the $Cl^- \rightarrow Cl_2$ oxidation, but other HCl does not undergo redox; it provides the H^+ ions for water and Cl^- ions for the KCl and $CrCl_3$.

Step 7 - Chlorine: 14 Cl atoms on the left and 12 Cl atoms on the right.

Add 2 Cl atoms to the right. From the original equation, the right-sided Cl atoms come in KCl; do not modify the Cl_2 since it has already been balanced by the oxidation-state method.

When balancing equations at this stage, only manipulate the water and spectator species.

$$K_2Cr_2O_7 + 14\ HCl \rightarrow 2\ CrCl_3 + 3\ Cl_2 + 7\ H_2O + \mathbf{2\ KCl}$$

Balanced redox equation is:

$$K_2Cr_2O_7\ (aq) + 14\ HCl\ (aq) \rightarrow 2\ CrCl_3\ (aq) + 3\ Cl_2\ (g) + 7\ H_2O\ (l) + 2\ KCl\ (aq)$$

Ten steps for balancing redox half-reactions

1. Split equation into two half-reactions.

2. Balance atoms other than O or H.

3. Balance O using H_2O. Add water as needed to balance the oxygen on the side deficient in oxygen.

4. If in an acidic solution, balance hydrogens using H^+.

    Add H^+ ions as needed to the side deficient in H.

5. In a basic solution, add OH^- ions equal to the number of H^+ ions to each side of the chemical equation.

6. The mass is balanced. To balance charge, determine the charge on each side of the reaction and add the necessary electrons to the positive side, so that the charges are equal.

7. Repeat steps 1 through 4 for each half-reaction.

8. If electrons lost do not equal the number gained, multiply each half-reaction by the necessary factor.

9. Sum half-reactions to obtain the balanced net ionic reaction. Inspect H^+ ions and H_2O molecules to cancel.

10. Check the final equation for mass and charge balance.

Important: If expressing H^+ as H_3O^+, change the number of H^+ into the same number of H_3O^+.

Add that same number of H_2O to the opposite side of the equation.

Limiting Reactant

Identifying limiting reactants

Limiting reactant is the component present in the lowest stoichiometric quantity, thereby limiting the product formed.

The limiting reactant is the reagent *depleted first and terminates the reaction.*

For example, identify the limiting reactant when a 50.6 g sample of magnesium hydroxide $Mg(OH)_2$ reacts with 45.0 g of hydrogen chloride, HCl, according to the reaction below:

$$Mg(OH)_2 + 2\ HCl \rightarrow MgCl_2 + 2\ H_2O$$

Notice that the quantities of each reactant are known.

There are three common methods of solving limiting reagent problems.

Moles to determine the limiting reactant

To identify the limiting reagent, *convert grams of reactants to moles* using molar mass:

$$50.6\ g\ Mg(OH)_2 \times (1\ mol\ Mg(OH)_2\ /\ 58.3\ g\ Mg(OH)_2) = 0.868\ mol\ Mg(OH)_2$$

$$45.0\ g\ HCl \times (1\ mol\ HCl\ /\ 36.5\ g\ HCl) = 1.23\ mol\ HCl$$

Select one reactant and calculate the moles of the other reactant needed to deplete the reactant.

For example, begin with magnesium hydroxide $(Mg(OH)_2)$:

$$0.868\ mol\ Mg(OH)_2 \times (2\ mol\ HCl\ needed/1\ mol\ Mg(OH)_2) = 1.74\ mol\ HCl\ needed$$

Compare the moles of HCl needed to the actual moles of HCl available.

In this example, 1.74 moles of HCl are needed, and 1.23 moles of HCl are present.

Even though HCl has more moles than $Mg(OH)_2$, HCl is the limiting reagent.

HCl is consumed before the $Mg(OH)_2$, limiting the product formed.

Theoretical to actual ratio to determine the limiting reactant

Theoretical yield is the amount of product expected from a reaction.

Compare *theoretical ratio* to the *actual ratio* of reactants.

Determine the moles of each reactant using molar mass:

$$50.6\ g\ Mg(OH)_2 \times (1\ mol\ Mg(OH)_2\ /\ 58.3\ g\ Mg(OH)_2) = 0.868\ mol\ Mg(OH)_2\ available$$

$$45.0\ g\ HCl \times (1\ mol\ HCl\ /\ 36.5\ g\ HCl) = 1.23\ mol\ HCl\ available$$

For example, consider the balanced reaction:

$$Mg(OH)_2 + 2\ HCl \rightarrow MgCl_2 + 2\ H_2O$$

From the balanced equation, the theoretical mole ratio is:

2 moles of HCl needed / 1 mol $Mg(OH)_2$

The actual mole ratio, based on the amounts of reactants present:

1.23 mol HCl / 0.868 $Mg(OH)_2$ = 1.42 mol HCl present / 1 mol $Mg(OH)_2$

There is not enough HCl from these ratios, so HCl is the limiting reagent.

2 mol HCl needed / 1 mol $Mg(OH)_2$ *vs.* **1.42 mol HCl present** / 1 mol $Mg(OH)_2$

Comparing theoretical yield to determine the limiting reactant

Calculate each *reactant's theoretical yield* and choose the lesser as the limiting reagent.

$$theoretical\ yield = 50.6\ g\ Mg(OH)_2 \times \frac{1\ mol\ Mg(OH)_2}{58.3\ g\ Mg(OH)_2} \times \frac{1\ mol\ MgCl_2}{1\ mol\ Mg(OH)_2} \times \frac{95.3\ g\ MgCl_2}{1\ mol\ MgCl_2}$$

theoretical yield = 82.7 *g* $MgCl_2$

$$theoretical\ yield = 45.0\ g\ HCl \times \frac{1\ mol\ HCl}{36.5\ g\ HCl} \times \frac{1\ mol\ MgCl_2}{2\ mol\ HCl} \times \frac{95.3\ g\ MgCl_2}{1\ mol\ MgCl_2}$$

theoretical yield = 58.6 *g* $MgCl_2$

HCl produced less product (i.e., limiting reagent); 58.6 g $MgCl_2$ as theoretical yield.

Theoretical yield and percent yield

Reactions often yield less product than predicted for several reasons (e.g., equipment inefficiency or human error).

Percent yield is the experimental (actual) ratio compared to the theoretical yield.

percent yield = (experimental yield / theoretical yield) × 100%

For example, if the reaction of 30.0 grams of calcium carbonate ($CaCO_3$) produces 15.0 grams of calcium oxide (CaO), what is the percent yield for the following reaction?

$$CaCO_3 \rightarrow CaO + CO_2$$

Experimental yield was 15.0 g, so the theoretical yield must be calculated.

For every mole of $CaCO_3$ reacted, one mole of CaO was produced.

From periodic table, molar mass of $CaCO_3$ is 100.0896, and the molar mass of CaO is 56.0774.

Use conversion factors to perform stoichiometric calculations and determine theoretical yield.

Theoretical yield = 30.0 g $CaCo_3$ × (1 mol $CaCO_3$ / 100.0869 *g*) × (1 mol CaO / 1 mol $CaCO_3$)

× (56.0774 *g* CaO / 1 mol CaO)

Theoretical yield = 16.8 *g* CaO

Use the formula above to calculate *percent yield*.

% yield = (actual yield / theoretical yield) × 100%

% yield = (15.0 *g* CaO / 16.8 *g* CaO) × 100%

% yield = 89.3%

Notes for active learning

Notes for active learning

Notes for active learning

PRACTICE QUESTIONS
&
DETAILED EXPLANATIONS

Practice Questions

Practice Set 1: Questions 1–20

1. What is the volume of three moles of O_2 at STP?

   A. 11.20 L

   B. 22.71 L

   C. 68.13 L

   D. 32.00 L

   E. 5.510 L

2. In the following reaction, which of the following describes H_2SO_4?

   $$H_2SO_4 + HI \rightarrow I_2 + SO_2 + H_2O$$

   A. reducing agent and is reduced

   B. reducing agent and is oxidized

   C. oxidizing agent and is reduced

   D. oxidizing agent and is oxidized

   E. neither an oxidizing nor reducing agent

3. Select the balanced chemical equation for the reaction: $C_6H_{14} + O_2 \rightarrow CO_2 + H_2O$

   A. $3\ C_6H_{14} + O_2 \rightarrow 18\ CO_2 + 22\ H_2O$

   B. $2\ C_6H_{14} + 12\ O_2 \rightarrow 12\ CO_2 + 14\ H_2O$

   C. $2\ C_6H_{14} + 19\ O_2 \rightarrow 12\ CO_2 + 14\ H_2O$

   D. $2\ C_6H_{14} + 9\ O_2 \rightarrow 12\ CO_2 + 7\ H_2O$

   E. $C_6H_{14} + O_2 \rightarrow CO_2 + H_2O$

4. An oxidation number is the [　] that an atom [　] when the electrons in each bond are assigned to the [　] electronegative of the two atoms involved in the bond.

   A. number of protons … definitely has … more

   B. charge … definitely has … less

   C. number of electrons … definitely has … more

   D. number of electrons … appears to have … less

   E. charge … appears to have … more

5. What is the empirical formula of acetic acid (CH_3COOH)?

    A. CH_3COOH **C.** CH_2O

    B. $C_2H_4O_2$ **D.** CO_2H_2

                                        E. CHO

6. In which of the following compounds does Cl have an oxidation number of +7?

    A. $NaClO_2$ **C.** $Ca(ClO_3)_2$

    B. $Al(ClO_4)_3$ **D.** $LiClO_3$

                                        E. none of the above

7. What is the formula mass of a molecule of CO_2?

    A. 44 amu   **C.** 56.5 amu

    B. 52 amu   **D.** 112 amu

                                        E. None of the above

8. What is the product of heating cadmium metal and powdered sulfur?

    A. CdS_2 **C.** CdS

    B. Cd_2S_3 **D.** Cd_2S

                                        E. Cd_3S_2

9. After balancing the following redox reaction, what is the coefficient of NaCl?

$$Cl_2\ (g) + NaI\ (aq) \rightarrow I_2\ (s) + NaCl\ (aq)$$

    A. 1   **C.** 3

    B. 2   **D.** 5

                                      E. none of the above

10. Which substance listed below is the strongest reducing agent, given the following *spontaneous* redox reaction?

$$FeCl_3\ (aq) + NaI\ (aq) \rightarrow I_2\ (s) + FeCl_2\ (aq) + NaCl\ (aq)$$

    A. $FeCl_2$ **B.** I_2 **C.** NaI   **D.** $FeCl_3$ **E.** NaCl

11. 14.5 moles of N_2 gas is mixed with 34 moles of H_2 gas in the following reaction:

$$N_2\,(g) + 3\,H_2\,(g) \rightarrow 2\,NH_3\,(g)$$

How many moles of N_2 gas remain if the reaction produces 18 moles of NH_3 gas when performed at 600 K?

A. 0.6 moles

B. 1.4 moles

C. 5.5 moles

D. 7.4 moles

E. 9.6 moles

12. Which equation is NOT correctly classified by the type of chemical reaction?

A. $AgNO_3 + NaCl \rightarrow AgCl + NaNO_3$ (double-replacement/non-redox)

B. $Cl_2 + F_2 \rightarrow 2\,ClF$ (synthesis/redox)

C. $H_2O + SO_2 \rightarrow H_2SO_3$ (synthesis/non-redox)

D. $CaCO_3 \rightarrow CaO + CO_2$ (decomposition/redox)

E. All are correctly classified

13. How many grams of H_2O can be formed from a reaction between 10 grams of oxygen and 1 gram of hydrogen?

A. 11 grams of H_2O are formed since mass must be conserved

B. 10 grams of H_2O are formed since the mass of water produced cannot be greater than the amount of oxygen reacting

C. 9 grams of H_2O are formed because oxygen and hydrogen react in an 8:1 mass ratio

D. No H_2O is formed because there is insufficient hydrogen to react with the oxygen

E. Not enough information is provided

14. How many grams of Ba^{2+} ions are in an aqueous solution of $BaCl_2$ that contains 6.8×10^{22} Cl ions?

A. 3.2×10^{48} g

B. 12 g

C. 14.5 g

D. 7.8 g

E. 9.8 g

15. What is the oxidation number of Br in $NaBrO_3$?

A. −1

B. +1

C. +3

D. +5

E. none of the above

16. When aluminum metal reacts with ferric oxide (Fe_2O_3), a displacement reaction yields two products, with one being metallic iron. What is the sum of the coefficients of the products of the balanced reaction?

A. 4 **B.** 6 **C.** 2 **D.** 5 **E.** 3

17. Which of the following reactions is NOT correctly classified?

A. $AgNO_3$ (*aq*) + KOH (*aq*) → KNO_3 (*aq*) + AgOH (*s*) : non-redox / double replacement
B. 2 H_2O_2 (*s*) → 2 H_2O (*l*) + O_2 (*g*) : non-redox / decomposition
C. $Pb(NO_3)_2$ (*aq*) + 2 Na (*s*) → Pb (*s*) + 2 $NaNO_3$ (*aq*) : redox / single-replacement
D. HNO_3 (*aq*) + LiOH (*aq*) → $LiNO_3$ (*aq*) + H_2O (*l*) : non-redox / double-replacement
E. All are correctly classified

18. Calculate the number of O_2 molecules if a 15.0 L cylinder was filled with O_2 gas at STP. (Use the conversion factor of 1 mole of O_2 = 6.02×10^{23} O_2 molecules)

A. 443 molecules
B. 6.59×10^{24} molecules
C. 4.03×10^{23} molecules
D. 2.77×10^{22} molecules
E. 4,430 molecules

19. Which of the following is a guideline for balancing redox equations by the oxidation number method?

A. Verify that the number of atoms and the ionic charge are the same for reactants and products
B. In front of the substance reduced, place a coefficient that corresponds to the number of electrons lost by the substance oxidized
C. In front of the substance oxidized, place a coefficient that corresponds to the number of electrons gained by the substance reduced
D. Determine the electrons lost by the substance oxidized and gained by the substance reduced
E. All of the above

20. Which of the following represents the oxidation of Co^{2+}?

A. Co → Co^{2+} + 2 e^-
B. Co^{3+} + e^- → Co^{2+}
C. Co^{2+} + 2 e^- → Co
D. Co^{2+} → Co^{3+} + e^-
E. Co^{3+} + 2 e^- → Co^+

Practice Set 2: Questions 21–40

21. How many moles of phosphorous trichloride are required to produce 365 grams of HCl when the reaction yields 75%?

$$PCl_3\ (g) + 3\ NH_3\ (g) \rightarrow P(NH_2)_3 + 3\ HCl\ (g)$$

A. 1 mol　　　**B.** 2.5 mol　　　**C.** 3.5 mol　　　**D.** 4.5 mol　　　**E.** 5 mol

22. What is the oxidation number of liquid bromine in the elemental state?

A. 0

B. –1

C. –2

D. –3

E. None of the above

23. Propane burners are used by campers for cooking. What volume of H_2O is produced by the complete combustion, as shown in the unbalanced equation, of 2.6 L of propane (C_3H_8) gas when measured at the same temperature and pressure?

$$C_3H_8\ (g) + O_2\ (g) \rightarrow CO_2\ (g) + H_2O\ (g)$$

A. 0.65 L　　　**B.** 10.4 L　　　**C.** 5.2 L　　　**D.** 2.6 L　　　**E.** 26.0 L

24. Which substance is reduced in the following redox reaction?

$$HgCl_2\ (aq) + Sn^{2+}\ (aq) \rightarrow Sn^{4+}\ (aq) + Hg_2Cl_2\ (s) + Cl^-\ (aq)$$

A. Sn^{4+}

B. Hg_2Cl_2

C. $HgCl_2$

D. Sn^{2+}

E. None of the above

25. What is the coefficient (n) of P for the balanced equation: $nP\ (s) + nO_2\ (g) \rightarrow nP_2O_5\ (s)$?

A. 1

B. 2

C. 4

D. 5

E. none of the above

26. In basic solution, which of the following are guidelines for balancing a redox equation by the half-reaction method?

I. Add the two half-reactions and cancel identical species on each side of the equation

II. Multiply each half-reaction by a whole number so that the number of electrons lost by the substance oxidized is equal to the electrons gained by the substance reduced

III. Write a balanced half-reaction for the substance oxidized and the substance reduced

A. II only

B. III only

C. I and III only

D. II and III only

E. I, II and III

27. What is the molecular formula of galactose if the empirical formula is CH_2O, and the approximate molar mass is 180 g/mol?

A. $C_6H_{12}O_6$

B. CH_2O_6

C. CH_2O

D. CHO

E. $C_{12}H_{22}O_{11}$

28. How many formula units of lithium iodide (LiI) have a mass equal to 6.45 g? (Use the molecular mass of LiI = 133.85 g)

A. 3.45×10^{23} formula units

B. 6.43×10^{23} formula units

C. 1.65×10^{24} formula units

D. 7.74×10^{25} formula units

E. 2.90×10^{22} formula units

29. What is/are the product(s) of the reaction of N_2 and O_2 gases in a combustion engine?

I. NO $\qquad$ II. NO_2 $\qquad$ III. N_2O

A. I only

B. II only

C. III only

D. I and II only

E. I, II and III

30. What is the coefficient (n) of O_2 gas for the balanced equation?

$$n\text{P } (s) + n\text{O}_2 (g) \rightarrow n\text{P}_2\text{O}_3 (s)$$

A. 1

B. 2

C. 3

D. 5

E. None of the above

31. Which substance is the weakest reducing agent given the spontaneous redox reaction?

$$Mg\ (s) + Sn^{2+}\ (aq) \rightarrow Mg^{2+}\ (aq) + Sn\ (s)$$

A. Sn

B. Mg^{2+}

C. Sn^{2+}

D. Mg

E. None of the above

32. After balancing the following redox reaction in acidic solution, what is the coefficient of H^+?

$$Mg\ (s) + NO_3^-\ (aq) \rightarrow Mg^{2+}\ (aq) + NO_2\ (aq)$$

A. 1

B. 2

C. 4

D. 6

E. None of the above

33. Which substance contains the greatest number of moles in a 10 g sample?

A. SiO_2 **B.** SO_2 **C.** CBr_4 **D.** CO_2 **E.** CH_4

34. Which chemistry law is illustrated when ethyl alcohol always contains 52% carbon, 13% hydrogen, and 35% oxygen by mass?

A. law of constant composition

B. law of constant percentages

C. law of multiple proportions

D. law of conservation of mass

E. none of the above

35. Which could NOT be true for the following reaction: $N_2\ (g) + 3\ H_2\ (g) \rightarrow 2\ NH_3\ (g)$?

A. 25 grams of N_2 gas reacts with 75 grams of H_2 gas to form 50 grams of NH_3 gas

B. 28 grams of N_2 gas reacts with 6 grams of H_2 gas to form 34 grams of NH_3 gas

C. 15 moles of N_2 gas reacts with 45 moles of H_2 gas to form 30 moles of NH_3 gas

D. 5 molecules of N_2 gas reacts with 15 molecules of H_2 gas, forming 10 molecules of NH_3 gas

E. None of the above

36. From the following reaction, if 0.2 moles of Al is allowed to react with 0.4 moles of Fe_2O_3, how many grams of aluminum oxide is produced?

$$2\,Al + Fe_2O_3 \rightarrow 2\,Fe + Al_2O_3$$

A. 2.8 g **B.** 5.1 g **C.** 10.2 g **D.** 14.2 g **E.** 18.6 g

37. In all of the following compounds, the oxidation number of hydrogen is +1, EXCEPT:

A. NH_3

B. $HClO_2$

C. H_2SO_4

D. NaH

E. none of the above

38. Which equation is NOT correctly classified by the type of chemical reaction?

A. $PbO + C \rightarrow Pb + CO$: single-replacement/non-redox

B. $2\,Na + 2HCl \rightarrow 2\,NaCl + H_2$: single-replacement/redox

C. $NaHCO_3 + HCl \rightarrow NaCl + H_2O + CO_2$: double-replacement/non-redox

D. $2\,Na + H_2 \rightarrow 2\,NaH$: synthesis/redox

E. All are correctly classified

39. What are the oxidation states of sulfur in H_2SO_4 and H_2SO_3, respectively?

A. +4 and +4

B. +6 and +4

C. +2 and +4

D. +4 and +2

E. +4 and +6

40. A latex balloon has a volume of 500 mL when filled with gas at a pressure of 780 torr and a temperature of 320 K. How many moles of gas does the balloon contain? (Use the ideal gas constant $R = 0.08206$ L·atm K^{-1} mol^{-1})

A. 0.0195 **B.** 0.822 **C.** 3.156 **D.** 18.87 **E.** 1.282

Practice Set 3: Questions 41–60

41. What is the term for a substance that causes the reduction of another substance in a redox reaction?

   A. oxidizing agent

   B. reducing agent

   C. anode

   D. cathode

   E. none of the above

42. Which of the following is a method for balancing a redox equation in acidic solution by the half-reaction method?

   A. Multiply each half-reaction by a whole number so that the number of electrons lost by the substance oxidized is equal to the electrons gained by the substance reduced

   B. Add the two half-reactions and cancel identical species from each side of the equation

   C. Write a half-reaction for the substance oxidized and the substance reduced

   D. Balance the atoms in each half-reaction; balance O with water and hydrogen with H^+

   E. All of the above

43. What is the formula mass of a molecule of $C_6H_{12}O_6$?

   A. 148 amu

   B. 27 amu

   C. 91 amu

   D. 180 amu

   E. None of the above

44. Is it possible to have a macroscopic sample of oxygen that has a mass of 12 atomic mass units?

   A. No, because oxygen is a gas at room temperature

   B. Yes, because it would have the same density as nitrogen

   C. No, because this is less than a macroscopic quantity

   D. Yes, but it would need to be made of isotopes of oxygen atoms

   E. No, because this is less than the mass of a single oxygen atom

45. What is the coefficient for O_2 when balanced with the lowest whole number coefficients?

$$C_2H_6 + O_2 \rightarrow CO_2 + H_2O$$

A. 3   **B.** 4   **C.** 6   **D.** 7   **E.** 9

46. Which element is reduced in the following redox reaction?

$$BaSO_4 + 4\,C \rightarrow BaS + 4\,CO$$

A. O in CO   **C.** S in BaS

B. Ba in BaS   **D.** C in CO

                                  E. S in $BaSO_4$

47. Ethanol (C_2H_5OH) is blended with gasoline as a fuel additive. If the combustion of ethanol produces carbon dioxide and water, what is the coefficient of oxygen in the balanced equation?

Spark

$$\_C_2H_5OH\,(g) + \_O_2\,(g) \rightarrow \_CO_2\,(g) + \_H_2O\,(g)$$

A. 1   **C.** 3

B. 2   **D.** 6

                                  E. None of the above

48. How many moles of C are in a 4.50 g sample if the atomic mass of C is 12.011 amu?

A. 5.40 moles   **C.** 1.00 moles

B. 2.67 moles   **D.** 0.54 moles

                                  E. 0.375 moles

49. Upon combustion analysis, a 6.84 g sample of a hydrocarbon yielded 8.98 grams of carbon dioxide. The percent, by mass, of carbon in the hydrocarbon is:

A. 18.6%   **B.** 23.7%   **C.** 35.8%   **D.** 11.4%   **E.** 52.8%

50. Select the balanced chemical equation:

A. $2\,C_2H_5OH + 2\,Na_2Cr_2O_7 + 8\,H_2SO_4 \rightarrow 2\,HC_2H_3O_2 + 2\,Cr_2(SO_4)_3 + 4\,Na_2SO_4 + 11\,H_2O$

B. $2\,C_2H_5OH + Na_2Cr_2O_7 + 8\,H_2SO_4 \rightarrow 3\,HC_2H_3O_2 + 2\,Cr_2(SO_4)_3 + 2\,Na_2SO_4 + 11\,H_2O$

C. $C_2H_5OH + 2\,Na_2Cr_2O_7 + 8\,H_2SO_4 \rightarrow HC_2H_3O_2 + 2\,Cr_2(SO_4)_3 + 2\,Na_2SO_4 + 11\,H_2O$

D. $C_2H_5OH + Na_2Cr_2O_7 + 2\,H_2SO_4 \rightarrow HC_2H_3O_2 + Cr_2(SO_4)_3 + 2\,Na_2SO_4 + 11\,H_2O$

E. $3\,C_2H_5OH + 2\,Na_2Cr_2O_7 + 8\,H_2SO_4 \rightarrow 3\,HC_2H_3O_2 + 2\,Cr_2(SO_4)_3 + 2\,Na_2SO_4 + 11\,H_2O$

51. Under acidic conditions, what is the sum of the coefficients in the balanced reaction below?

$$Fe^{2+} + Cr_2O_7^{2-} \rightarrow Fe^{3+} + Cr^{3+}$$

A. 8

B. 14

C. 17

D. 36

E. none of the above

52. Assuming STP, if 49 g of H_2SO_4 are produced in the following reaction, what volume of O_2 must be used in the reaction?

$$RuS\ (s) + O_2 + H_2O \rightarrow Ru_2O_3\ (s) + H_2SO_4$$

A. 20.6 liters

B. 31.2 liters

C. 28.3 liters

D. 29.1 liters

E. 25.2 liters

53. From the following reaction, if 0.20 mole of Al is allowed to react with 0.40 mole of Fe_2O_3, how many moles of iron are produced?

$$2\ Al + Fe_2O_3 \rightarrow 2\ Fe + Al_2O_3$$

A. 0.05 mole

B. 0.075 mole

C. 0.20 mole

D. 0.10 mole

E. 0.25 mole

54. What is the oxidation number of sulfur in the $S_2O_8^{2-}$ ion?

A. −1   B. +7   C. +2   D. +6   E. +1

55. What are the oxidation numbers for the elements in Na_2CrO_4?

A. +2 for Na, +5 for Cr and −6 for O

B. +2 for Na, +3 for Cr and −2 for O

C. +1 for Na, +4 for Cr and −6 for O

D. +1 for Na, +6 for Cr and −2 for O

E. +1 for Na, +5 for Cr and −2 for O

56. What is the oxidation state of sulfur in sulfuric acid?

A. +8   B. +6   C. −2   D. −6   E. +4

57. Which substance is oxidized in the following redox reaction?

$$HgCl_2 \ (aq) + Sn^{2+} \ (aq) \rightarrow Sn^{4+} \ (aq) + Hg_2Cl_2 \ (s) + Cl^- \ (aq)$$

A. Hg_2Cl_2

B. Sn^{4+}

C. Sn^{2+}

D. $HgCl_2$

E. None of the above

58. Which of the following statements that is NOT true with respect to the balanced equation?

$$Na_2SO_4 \ (aq) + BaCl_2 \ (aq) \rightarrow 2 \ NaCl \ (aq) + BaSO_4 \ (s)$$

A. Barium sulfate and sodium chloride are products

B. Barium chloride is dissolved in water

C. Barium sulfate is a solid

D. $2 \ NaCl \ (aq)$ could be written as $Na_2Cl_2 \ (aq)$

E. Sodium sulfate has a coefficient of one

59. Which of the following represents 1 mol of phosphine gas (PH_3)?

    I. 22.71 L phosphine gas at STP

    II. 34.00 g phosphine gas

    III. 6.02×10^{23} phosphine molecules

A. I only

B. II only

C. III only

D. I and II only

E. II and III only

60. Which of the following represents the oxidation of Co^{2+}?

A. $Co \rightarrow Co^{2+} + 2 \ e^-$

B. $Co^{3+} + e^- \rightarrow Co^{2+}$

C. $Co^{2+} + 2 \ e^- \rightarrow Co$

D. $Co^{2+} \rightarrow Co^{3+} + e^-$

E. $Co^{3+} + 2 \ e^- \rightarrow Co^+$

Practice Set 4: Questions 61–80

61. What is the term for a substance that causes oxidation in a redox reaction?

A. oxidized

B. reducing agent

C. anode

D. cathode

E. oxidizing agent

62. Carbon tetrachloride (CCl_4) is a potent hepatotoxin (toxic to the liver) commonly used in the past in fire extinguishers and as a refrigerant. What is the percent by mass of Cl in carbon tetrachloride?

A. 25%   **B.** 66%   **C.** 78%   **D.** 92%   **E.** 33%

63. For a redox reaction to be balanced, which of the following is true?

I. Ionic charge of reactants must equal ionic charge of products

II. Atoms of each reactant must equal atoms of the product

III. Electron gain must equal electron loss

A. I only

B. II only

C. I and II only

D. I and III only

E. I, II and III

64. The mass percent of a compound is approximately 71.8% Cl, 24.2% C, and 4.0% H. If the molecular weight of the compound is 99 g/mol, what is the molecular formula of the compound?

A. $Cl_2C_3H_6$

B. $Cl_2C_2H_4$

C. $ClCH_3$

D. ClC_2H_2

E. Cl_3CH_3

65. In the early 1980s, benzene was used as a solvent for waxes and oils but is now listed as a carcinogen by the EPA. What is the molecular formula of benzene if the empirical formula is C_1H_1 and the approximate molar mass is 78 g/mol?

A. CH_{12} **B.** $C_{12}H_{12}$ **C.** CH   **D.** CH_6 **E.** C_6H_6

66. How many O atoms are in the formula unit $GaO(NO_3)_2$?

   A. 3   **B.** 4   **C.** 5   **D.** 7   **E.** 8

67. Which reaction represents the balanced reaction for the combustion of ethanol?

   A. $4\ C_2H_5OH + 13\ O_2 \rightarrow 8\ CO_2 + 10\ H_2$

   B. $C_2H_5OH + 3\ O_2 \rightarrow 2\ CO_2 + 3\ H_2O$

   C. $C_2H_5OH + 2\ O_2 \rightarrow 2CO_2 + 2\ H_2O$

   D. $C_2H_5OH + O_2 \rightarrow CO_2 + H_2O$

   E. $C_2H_5OH + \frac{1}{2}\ O_2 \rightarrow 2\ CO_2 + 3\ H_2O$

68. What is the coefficient for O_2 when the following equation is balanced with the lowest whole number coefficients?

$$\_\_C_3H_7OH + \_\_O_2 \rightarrow \_\_CO_2 + \_\_H_2O$$

   A. 3   **B.** 6   **C.** 9   **D.** 13/2   **E.** 12

69. After balancing the following redox reaction, what is the coefficient of CO_2?

$$\_\_Co_2O_3\ (s) + \_\_CO\ (g) \rightarrow \_\_Co\ (s) + \_\_CO_2\ (g)$$

   A. 1   **B.** 2   **C.** 3   **D.** 4   **E.** 5

70. What is the term for the volume occupied by 1 mol of any gas at STP?

   A. STP volume   **C.** standard volume

   B. molar volume   **D.** Avogadro's volume

                                E. none of the above

71. Which substance listed below is the weakest oxidizing agent given the following spontaneous redox reaction?

$$Mg\ (s) + Sn^{2+}\ (aq) \rightarrow Mg^{2+}\ (aq) + Sn\ (s)$$

   A. Mg^{2+} **C.** Mg

   B. Sn   **D.** Sn^{2+}

                           E. none of the above

72. In the following reaction performed at 500 K, 18.0 moles of N_2 gas is mixed with 24.0 moles of H_2 gas. What is the percent yield of NH_3 if the reaction produces 13.5 moles of NH_3?

$$N_2\,(g) + 3\,H_2\,(g) \rightarrow 2\,NH_3\,(g)$$

A. 16%   **B.** 66%   **C.** 72%   **D.** 84%   **E.** 100%

73. How many grams of H_2O can be produced from the reaction of 25.0 grams of H_2 and 225 grams of O_2?

A. 266 grams

B. 223 grams

C. 184 grams

D. 27 grams

E. 2.5 grams

74. What is the oxidation number of Cl in $LiClO_2$?

A. −1

B. +1

C. +3

D. +5

E. None of the above

75. In acidic conditions, what is the sum of the coefficients in the products of the balanced reaction?

$$MnO_4^- + C_3H_7OH \rightarrow Mn^{2+} + C_2H_5COOH$$

A. 12   **B.** 16   **C.** 18   **D.** 20   **E.** 24

76. Which reaction is NOT correctly classified?

A. $PbO\,(s) + C\,(s) \rightarrow Pb\,(s) + CO\,(g)$: (double-replacement)

B. $CaO\,(s) + H_2O\,(l) \rightarrow Ca(OH)_2\,(aq)$: (synthesis)

C. $Pb(NO_3)_2\,(aq) + 2LiCl\,(aq) \rightarrow 2\,LiNO_3\,(aq) + PbCl_2\,(s)$: (double-replacement)

D. $Mg\,(s) + 2\,HCl\,(aq) \rightarrow MgCl_2\,(aq) + H_2\,(g)$: (single-replacement)

E. All are classified correctly

77. The reactants for this chemical reaction are:

$$C_6H_{12}O_6 + 6\,H_2O + 6\,O_2 \rightarrow 6\,CO_2 + 12\,H_2O$$

A. $C_6H_{12}O_6$, H_2O, O_2 and CO_2

B. $C_6H_{12}O_6$ and H_2O

C. $C_6H_{12}O_6$

D. $C_6H_{12}O_6$ and CO_2

E. $C_6H_{12}O_6$, H_2O and O_2

78. What is the charge of the electrons in 4 grams of He? (Use Faraday constant $F = 96{,}500$ C/mol)

A. 48,250 C

B. 96,500 C

C. 193,000 C

D. 386,000 C

E. Cannot be determined

79. What is the oxidation number of iron in the compound $FeBr_3$?

A. –2   B. +1   C. +2   D. +3   E. –1

80. What is the term for the amount of substance that contains 6.02×10^{23} particles?

A. molar mass

B. mole

C. Avogadro's number

D. formula mass

E. none of the above

Practice Set 5: Questions 81–100

81. Which of the following species undergoes oxidation in $2\,CuBr \rightarrow 2\,Cu + Br_2$?

    A. Cu^+ **B.** Br^- **C.** Cu **D.** $CuBr$ **E.** Br_2

82. How many electrons are lost or gained by each formula unit of $CuBr_2$ in this reaction?

$$Zn + CuBr_2 \rightarrow ZnBr_2 + Cu$$

    A. loses 1 electron   **C.** gains 2 electrons

    B. gains 6 electrons   **D.** loses 2 electrons

                             E. gains 4 electrons

83. How many molecules of CO_2 are in 168.0 grams of CO_2?

    A. 3.96×10^{23} **C.** 4.24×10^{22}

    B. 2.30×10^{24} **D.** 6.82×10^{24}

                             E. 4.60×10^{23}

84. What is the empirical formula of a compound that, by mass, contains 64% silver, 8% nitrogen and 28% oxygen?

    A. Ag_3NO **C.** $AgNO_2$

    B. Ag_3NO_3 **D.** Ag_3NO_2

                             E. $AgNO_3$

85. What is the mass of one mole of a gas that has a density of 1.34 g/L at STP?

    A. 48.0 g   **C.** 18.3 g

    B. 56.4 g   **D.** 30.1 g

                             E. 4.39 g

86. Propane (C_3H_8) is flammable and used as a substitute for natural gas. What is the coefficient of oxygen in the balanced equation for the combustion of propane?

$$\text{Spark}$$
$$\_\_C_3H_8\ (g) + \_\_O_2\ (g) \rightarrow \_\_CO_2\ (g) + \_\_H_2O\ (g)$$

A. 1

B. 7

C. 5

D. 10

E. None of the above

87. This reaction yields how many atoms of oxygen?

$$C_6H_{12}O_6 + 6\ H_2O + 6\ O_2 \rightarrow 6\ CO_2 + 12\ H_2O$$

A. 3   **B.** 12   **C.** 14   **D.** 24   **E.** 36

88. After balancing the following redox reaction, what is the coefficient of O_2?

$$\_\_Al_2O_3\ (s) + \_\_Cl_2\ (g) \rightarrow \_\_AlCl_3\ (aq) + \_\_O_2\ (g)$$

A. 1

B. 2

C. 3

D. 5

E. none of the above

89. Which reaction is the correctly balanced half-reaction (in acid solution) for the process below?

$$Cr_2O_7{}^{2-}\ (aq) \rightarrow Cr^{3+}\ (aq)$$

A. $8\ H^+ + Cr_2O_7 \rightarrow 2\ Cr^{3+} + 4\ H_2O + 3\ e^-$

B. $12\ H^+ + Cr_2O_7{}^{2-} + 3\ e^- \rightarrow 2\ Cr^{3+} + 6\ H_2O$

C. $14\ H^+ + Cr_2O_7{}^{2-} + 6\ e^- \rightarrow 2\ Cr^{3+} + 7\ H_2O$

D. $8\ H^+ + Cr_2O_7 + 3\ e^- \rightarrow 2\ Cr^{3+} + 4\ H_2O$

E. None of the above

90. Determine the oxidation number of C in $NaHCO_3$:

A. +5   **B.** +4   **C.** +12   **D.** +6   **E.** +2

91. What principle states that equal volumes of gases at the same temperature and pressure contain equal numbers of molecules?

A. Dalton's theory

B. Charles' theory

C. Boyle's theory

D. Avogadro's theory

E. None of the above

92. What are the products for this double-replacement reaction?

$$BaCl_2\ (aq) + K_2SO_4\ (aq) \rightarrow$$

A. $BaSO_3$ and $KClO_4$

B. $BaSO_4$ and $2\ KCl$

C. BaS and $KClO_4$

D. $BaSO_3$ and KCl

E. $BaSO_4$ and $KClO_4$

93. The formula for mustard gas used in chemical warfare is $C_4H_8SCl_2$. What is the percent by mass of chlorine in mustard gas? (Use the molecular mass of mustard gas = 159.09 g/mol)

A. 18.4%

B. 44.6%

C. 14.6%

D. 31.2%

E. 28.8%

94. Which substance listed is the weakest oxidizing agent given the following *spontaneous* redox reaction?

$$FeCl_3\ (aq) + NaI\ (aq) \rightarrow I_2\ (s) + FeCl_2\ (aq) + NaCl\ (aq)$$

A. I_2

B. $FeCl_2$

C. $FeCl_3$

D. NaI

E. $NaCl$

95. If the reaction below is run at STP with excess H_2O, and 22.4 liters of O_2 reacts with 67 g of RuS, how many grams of H_2SO_4 are produced?

$$RuS\ (s) + O_2 + H_2O \rightarrow Ru_2O_3\ (s) + H_2SO_4$$

A. 28 g **B.** 32 g **C.** 44 g **D.** 54 g **E.** 58 g

96. What is the oxidation number of sulfur in calcium sulfate, $CaSO_4$?

A. +6 **B.** +4 **C.** +2 **D.** 0 **E.** −2

97. How many moles of O_2 gas are required for combustion with one mole of $C_6H_{12}O_6$ in the unbalanced reaction?

$$C_6H_{12}O_6\ (s) + O_2\ (g) \rightarrow CO_2\ (g) + H_2O\ (g)$$

A. 1   **B.** 2.5   **C.** 6   **D.** 10   **E.** 12

98. Which reaction below is a *synthesis* reaction?

A. $3\ CuSO_4 + 2\ Al \rightarrow Al_2(SO_4)_3 + 3\ Cu$

B. $SO_3 + H_2O \rightarrow H_2SO_4$

C. $2\ NaHCO_3 \rightarrow Na_2CO_3 + CO_2 + H_2O$

D. $C_3H_8 + 5\ O_2 \rightarrow 3\ CO_2 + 4\ H_2O$

E. None of the above

99. Which statement is true regarding the coefficients in a chemical equation?

    I. They appear before the chemical formulas

    II. They appear as subscripts

    III. Coefficients of reactants sum to those of the products

A. I only

B. II only

C. I and II only

D. I and III only

E. I, II and III

100. What volume is occupied by 17.0 g of NO gas at STP?

A. 23.4 L   **B.** 58.4 L   **C.** 0.58 L   **D.** 12. 7 L   **E.** 46.8 L

Practice Set 6: Questions 101–120

101. Which of the following is/are likely to act as an oxidizing agent?

   I. Cl^- II. Cl_2 III. Na^+

   A. I only
   B. II only
   C. III only
   D. I and II only
   E. I, II and III

102. When a substance loses electrons, it is [   ], while the substance itself acts as [   ] agent.

   A. reduced… a reducing
   B. oxidized… a reducing
   C. reduced… an oxidizing
   D. oxidized… an oxidizing
   E. dissolved… a neutralizing

103. What is the term for the chemical formula of a compound that expresses the actual number of atoms of each element in a molecule?

   A. molecular formula
   B. empirical formula
   C. elemental formula
   D. atomic formula
   E. none of the above

104. Which substance is the weakest reducing agent given the spontaneous redox reaction?

   $$FeCl_3\ (aq) + NaI\ (aq) \rightarrow I_2\ (s) + FeCl_2\ (aq) + NaCl\ (aq)$$

   A. NaCl   **B.** I_2 **C.** NaI   **D.** $FeCl_3$ **E.** $FeCl_2$

105. What is the mass of 3.5 moles of glucose which has a molecular formula of $C_6H_{12}O_6$?

   A. 180 g
   B. 90 g
   C. 630 g
   D. 51.4 g
   E. 520 g

106. Butane (C_4H_{10}) is flammable and used in butane lighters. What is the coefficient of oxygen in the balanced equation for the combustion of butane?

$$\underset{\text{Spark}}{\_\_C_4H_{10}\ (g) + \_\_O_2\ (g) \rightarrow \_\_CO_2\ (g) + \_\_H_2O\ (g)}$$

A. 9

B. 13

C. 18

D. 26

E. none of the above

107. Which of the following is an accurate statement for Avogadro's number?

    I. Equal to the number of atoms in 1 mole of an element

    II. Equal to the number of molecules in a compound

    III. Equal to approximately 6.022×10^{23}

A. I only

B. II only

C. III only

D. I and III only

E. I, II and III

108. What is the sum of the coefficients in the balanced reaction (no fractional coefficients)?

$$\_\_RuS\ (s) + \_\_O_2 + \_\_H_2O \rightarrow \_\_Ru_2O_3\ (s) + \_\_H_2SO_4$$

A. 23

B. 18

C. 13

D. 25

E. 29

109. Which is sufficient for determining the molecular formula of a compound?

    I. The % by mass of a compound

    II. The molecular weight of a compound

    III. The amount of a compound in a sample

A. I only

B. II only

C. I and II only

D. I and III only

E. I, II and III

110. Which equation(s) is/are balanced?

    I. Mg (s) + 2 HCl (aq) $\rightarrow$ $MgCl_2$ (aq) + H_2 (g)

    II. 3 Al (s) + 3 Br_2 (l) $\rightarrow$ Al_2Br_3 (s)

    III. 2 HgO (s) $\rightarrow$ 2 Hg (l) + O_2 (g)

A. I only

B. II only

C. III only

D. I and II only

E. I and III only

111. How many moles are in 60.2 g of $MgSO_4$?

A. 0.25 mole

B. 0.5 mole

C. 0.45 mole

D. 0.65 mole

E. 0.75 mole

112. How many atoms of Mg are in a solid 72.9 g sample of magnesium? (Use the molecular mass of Mg = 24.3 g/mol)

A. 1.81×10^{24} atoms

B. 4,800 atoms

C. 3.01×10^{23} atoms

D. 2.42×10^{25} atoms

E. 6.02×10^{23} atoms

113. The formula for the illegal drug cocaine is $C_{17}H_{21}NO_4$ (303.39 g/mol). What is the percent by mass of oxygen in cocaine?

A. 21.1%   **B.** 6.5%   **C.** 457%   **D.** 4.4%   **E.** 62.8%

114. If 7 moles of RuS are used in the following reaction, what is the maximum number of moles of Ru_2O_3 that can be produced?

RuS (s) + O_2 + H_2O $\rightarrow$ Ru_2O_3 (s) + H_2SO_4

A. 3.5 moles

B. 1.8 moles

C. 5.8 moles

D. 2.2 moles

E. 1.6 moles

115. In which sequence are sulfur-containing species arranged to decrease oxidation numbers for S?

A. SO_4^{2-}, S^{2-}, $S_2O_3^{2-}$

B. $S_2O_3^{2-}$, SO_3^{2-}, S^{2-}

C. SO_4^{2-}, $S_2O_3^{2-}$, S^{2-}

D. SO_3^{2-}, SO_4^{2-}, S^{2-}

E. S^{2-}, SO_4^{2-}, $S_2O_3^{2-}$

116. In the following compounds, the oxidation number of oxygen is −2, EXCEPT:

A. $NaClO_2$

B. Li_2O_2

C. $Ba(OH)_2$

D. Na_2SO_4

E. none of the above

117. Which of the following is a *double-replacement* reaction?

A. $2\,HI \rightarrow H_2 + I_2$

B. $SO_2 + H_2O \rightarrow H_2SO_3$

C. $HBr + KOH \rightarrow H_2O + KBr$

D. $CuO + H_2 \rightarrow Cu + H_2O$

E. none of the above

118. Which of the following substances has the lowest density?

A. A mass of 750 g and a volume of 70 dL

B. A mass of 5 μg and a volume of 25 μL

C. A mass of 1.5 kg and a volume of 1.2 L

D. A mass of 25 g and a volume of 20 mL

E. A mass of 15 mg and a volume of 50 μL

119. How many moles of O_2 gas is required for combustion with 2 moles of hexane in the unbalanced reaction:

$$C_6H_{14}\,(g) + O_2\,(g) \rightarrow CO_2\,(g) + H_2O\,(g)$$

A. 11   **B.** 14   **C.** 12   **D.** 20   **E.** 19

120. How is Avogadro's number related to the numbers on the periodic table?

A. The atomic mass listed is the mass of Avogadro's number of atoms

B. The periodic table provides the mass of one atom, while Avogadro's number provides the number of moles

C. The masses are divisible by Avogadro's number, which provides the weight of one mole

D. The periodic table only provides atomic numbers, not atomic mass

E. The mass listed is in the units of Avogadro's number

Answer Key & Detailed Explanations

Answer Key

1: C	21: D	41: B	61: E	81: B	101: B
2: C	22: A	42: E	62: D	82: C	102: B
3: C	23: B	43: D	63: E	83: B	103: A
4: E	24: C	44: E	64: B	84: E	104: E
5: C	25: C	45: D	65: E	85: D	105: C
6: B	26: E	46: E	66: D	86: C	106: B
7: A	27: A	47: C	67: B	87: D	107: D
8: C	28: E	48: E	68: C	88: C	108: A
9: B	29: E	49: C	69: C	89: C	109: C
10: C	30: C	50: E	70: B	90: B	110: E
11: C	31: A	51: D	71: A	91: D	111: B
12: D	32: C	52: E	72: D	92: B	112: A
13: C	33: E	53: C	73: B	93: B	113: A
14: D	34: A	54: B	74: C	94: A	114: A
15: D	35: A	55: D	75: D	95: C	115: C
16: E	36: C	56: B	76: A	96: A	116: B
17: B	37: D	57: C	77: E	97: C	117: C
18: C	38: A	58: D	78: C	98: B	118: B
19: E	39: B	59: E	79: D	99: A	119: E
20: D	40: A	60: D	80: B	100: D	120: A

Practice Set 1: Questions 1–20

---

1. C is correct.

One mole of an ideal gas occupies 22.71 L at STP.

$$3 \text{ mole} \times 22.71 \text{ L} = 68.13 \text{ L}$$

2. C is correct.

Use the mnemonic OIL RIG: <u>O</u>xidation <u>I</u>s <u>L</u>oss, <u>R</u>eduction <u>I</u>s <u>G</u>ain (of electrons).

Oxidation is the loss of electrons, while reduction is the gain of electrons.

An oxidizing agent undergoes reduction, while a reducing agent undergoes oxidation.

Check the oxidation numbers of each atom.

The oxidation number for S is +6 as a reactant and +4 as a product, reducing it.

If a substance is reduced in a redox reaction, it is the oxidizing agent because it causes the other reagent (HI) to be oxidized.

3. C is correct.

Balanced equation (combustion): $2\ C_6H_{14} + 19\ O_2 \rightarrow 12\ CO_2 + 14\ H_2O$

4. E is correct.

An oxidation number is a charge on an atom that is not present in the elemental state when the element is neutral.

Electronegativity refers to the attraction an atom has for additional electrons.

5. C is correct.

CH_3COOH has 2 carbons, 4 hydrogens, and 2 oxygens: $C_2O_2H_4$

Reduce to the smallest coefficients by dividing by two: COH_2

6. B is correct.

Oxidation numbers of Cl in each compound:

$NaClO_2$	+3
$Al(ClO_4)_3$	+7
$Ca(ClO_3)_2$	+5
$LiClO_3$	+5

7. A is correct.

Formula mass is a synonym for molecular mass/molecular weight (MW).

Molecular mass = (atomic mass of C) + (2 × atomic mass of O)

Molecular mass = (12.01 g/mole) + (2 × 16.00 g/mole)

Molecular mass = 44.01 g/mole

Note: 1 amu = 1 g/mole

8. C is correct.

Cadmium is located in group II, so its oxidation number is +2.

Sulfur tends to form a sulfide ion (oxidation number –2).

The net oxidation number is zero, and therefore the compound is CdS.

9. B is correct.

Balanced reaction (single replacement):

$$Cl_2\ (g) + 2\ NaI\ (aq) \rightarrow I_2\ (s) + 2\ NaCl\ (aq)$$

10. C is correct.

Use the mnemonic OIL RIG: <u>O</u>xidation <u>Is</u> <u>L</u>oss, <u>R</u>eduction <u>Is</u> <u>G</u>ain (of electrons).

Oxidation is the loss of electrons, while reduction is the gain of electrons.

An oxidizing agent undergoes reduction, while a reducing agent undergoes oxidation.

When there are 2 reactants in a redox reaction, one will be oxidized and the other reduced. The oxidized species is the reducing agent and vice versa.

Determine the oxidation number of all atoms:

I changes its oxidation state from –1 as a reactant to 0 as a product.

I is oxidized; therefore, the compound that contains I (NaI) is the reducing agent. Species in their elemental state have an oxidation number of 0.

Since the reaction is spontaneous, NaI is the *strongest* reducing agent because it can reduce another species spontaneously without needing any external energy for the reaction to proceed.

11. C is correct.

To determine the remainder of a reactant, calculate how many moles of that reactant are required to produce the specified amount of product.

Use the coefficients to calculate the moles of the reactant:

$$\text{moles } N_2 = (1/2) \times 18 \text{ moles} = 9 \text{ moles } N_2$$

Therefore, the remainder is: $(14.5 \text{ moles } N_2) - (9 \text{ moles } N_2) = 5.5 \text{ moles } N_2$ remaining

12. D is correct.

$CaCO_3 \rightarrow CaO + CO_2$ is a decomposition reaction because one complex compound is being broken down into two or more parts. However, the redox part is incorrect because the reactants and products are compounds (no single elements), and there is no change in the oxidation state (i.e., no gain or loss of electrons).

A: $AgNO_3 + NaCl \rightarrow AgCl + NaNO_3$ is correctly classified as a double-replacement reaction (sometimes referred to as double-displacement) because parts of two ionic compounds are exchanged to make two new compounds. It is a non-redox reaction because the reactants and products are compounds (no single elements), and there is no change in oxidation state (i.e., no gain or loss of electrons).

B: $Cl_2 + F_2 \rightarrow 2 \text{ ClF}$ is correctly classified as a synthesis reaction because two species are combining to form a more complex chemical compound as the product. It is a redox reaction because the F atoms in F_2 are reduced (i.e., gain electrons), and the Cl atoms in Cl_2 are oxidized (i.e., lose electrons), forming a covalent compound.

C: $H_2O + SO_2 \rightarrow H_2SO_3$ is correctly classified as a synthesis reaction because two species are combining to form a more complex chemical compound as the product. It is a non-redox reaction because the reactants and products are compounds (no single elements), and there is no change in oxidation state (i.e., no gain or loss of electrons).

13. C is correct.

Hydrogen is being oxidized (or ignited) to produce water.

The balanced equation that describes the ignition of hydrogen gas in the air to produce water is:

$$\tfrac{1}{2} O_2 + H_2 \rightarrow H_2O$$

This equation suggests that for every mole of water produced in the reaction, 1 mole of hydrogen and a ½ mole of oxygen gas is needed.

continued...

The amount of limiting reactant available determines the maximum amount of water produced in the reaction. This requires identifying which reactant is the limiting reactant and can be done by comparing the number of moles of oxygen and hydrogen.

There are 10 grams of oxygen gas and 1 gram of hydrogen gas from the question. This is equivalent to 0.3125 moles of oxygen and 0.5 moles of hydrogen.

Because the reaction requires twice as much hydrogen as oxygen gas, the limiting reactant is hydrogen gas (0.3215 moles of O_2 requires 0.625 moles of hydrogen, but only 0.5 moles of H_2 are available).

Since one equivalent of water is produced for every equivalent of hydrogen that is burned, the amount of water produced is:

$$(18 \text{ grams/mol } H_2O) \times (0.5 \text{ mol } H_2O) = 9 \text{ grams of } H_2O$$

14. D is correct.

Calculate the moles of Cl^- ion:

Moles of Cl^- ion = number of Cl^- atoms / Avogadro's number

Moles of Cl^- ion = 6.8×10^{22} atoms / 6.02×10^{23} atoms/mole

Moles of Cl^- ion = 0.113 moles

In a solution, $BaCl_2$ dissociates:

$$BaCl_2 \rightarrow Ba^{2+} + 2\ Cl^-$$

The moles of from Cl^- previous calculation can be used to calculate the moles of Ba^{2+}:

Moles of Ba^{2+} = (coefficient of Ba^{2+} / coefficient Cl^-) × moles of Cl^-

Moles of Ba^{2+} = (½) × 0.113 moles

Moles of Ba^{2+} = 0.0565 moles

Calculate the mass of Ba^{2+}:

Mass of Ba^{2+} = moles of Ba^{2+} × atomic mass of Ba

Mass of Ba^{2+} = 0.0565 moles × 137.33 g/mole

Mass of Ba^{2+} = 7.76 g

15. D is correct.

Na	Br	O_3
+1	x	(3×-2)
+1	x	-6

The sum of charges in a neutral molecule is zero:

1 + oxidation number of Br:

$$1 + x + (-6) = 0$$

$$-5 + x = 0$$

$$x = +5$$

oxidation number of Br = +5

16. E is correct.

Balanced equation:

$$2\ Al + Fe_2O_3 \rightarrow 2\ Fe + Al_2O_3$$

The sum of the coefficients of the products = 3.

17. B is correct.

$2\ H_2O_2\ (s) \rightarrow 2\ H_2O\ (l) + O_2\ (g)$ is incorrectly classified. The decomposition part of the classification is correct because one complex compound (H_2O_2) is being broken down into two or more parts (H_2O and O_2).

However, it is a redox reaction because oxygen is being lost from H_2O_2, one of the three indicators of a redox reaction (i.e., electron loss/gain, hydrogen loss/gain, oxygen loss/gain). Additionally, one of the products is a single element (O_2), which is an indication that it is a redox reaction.

$AgNO_3\ (aq) + KOH\ (aq) \rightarrow KNO_3\ (aq) + AgOH\ (s)$ is correctly classified as a non-redox reaction because the reactants and products are compounds (no single elements), and there is no change in oxidation state (i.e., no gain or loss of electrons). It is also a precipitation reaction because the chemical reaction occurs in an aqueous solution, and one of the products formed (AgOH) is insoluble, which makes it a precipitate.

$Pb(NO_3)_2\ (aq) + 2\ Na\ (s) \rightarrow Pb\ (s) + 2\ NaNO_3\ (aq)$ is correctly classified as a redox reaction because the nitrate (NO_3^-), which dissociates, is oxidized (loses electrons), the Na^+ is reduced (gains electrons). The two combine to form the ionic compound $NaNO_3$. It is a single-replacement reaction because one element is substituted for another element in a compound, making a new compound ($2NaNO_3$) and an element (Pb). *continued...*

HNO_3 (*aq*) + LiOH (*aq*) $\rightarrow$ $LiNO_3$ (*aq*) + H_2O (*l*) is correctly classified as a non-redox reaction because the reactants and products are compounds (no single elements), and there is no change in oxidation state (i.e., no gain or loss of electrons). It is a double-replacement reaction because parts of two ionic compounds are exchanged to make two new compounds.

18. C is correct.

At STP, 1 mole of gas has a volume of 22.4 L.

Calculate moles of O_2:

moles of O_2 = (15.0 L) / (22.4 L/mole)

moles of O_2 = 0.67 mole

Calculate the moles of O_2, using the fact that 1 mole of O_2 has 6.02×10^{23} O_2 molecules:

molecules of O_2 = 0.67 mole $\times$ (6.02×10^{23} molecules/mole)

molecules of O_2 = 4.03×10^{23} molecules

19. E is correct.

20. D is correct.

Use the mnemonic OIL RIG: <u>O</u>xidation <u>I</u>s <u>L</u>oss, <u>R</u>eduction <u>I</u>s <u>G</u>ain (of electrons).

Oxidation is the loss of electrons, while reduction is the gain of electrons.

An oxidizing agent undergoes reduction, while a reducing agent undergoes oxidation.

Co^{2+} is the starting reactant because it has to lose electrons and produce an ion with a higher oxidation number.

Notes for active learning

Practice Set 2: Questions 21–40

21. D is correct.

Double replacement reaction:

HCl has a molar mass of 36.5 g/mol:

> 365 grams HCl = 10 moles HCl

> 10 moles are 75% of the yield.

> 10 moles PCl_3 / 0.75 = 13.5 moles of HCl.

Apply the mole ratios:

> Coefficients: 3 HCl = PCl_3

> 13.5 / 3 = moles of PCl_3

> 13.5 moles of HCl is produced by 4.5 moles of PCl_3

22. A is correct.

All elements in their free state have an oxidation number of zero.

23. B is correct.

Balanced chemical equation:

> $C_3H_8 + 5\,O_2 \rightarrow 3\,CO_2 + 4\,H_2O$

By using the coefficients from the balanced equation, apply dimensional analysis to solve for the volume of H_2O produced from the 2.6 L C_3H_8 reacted:

> $V_{H2O} = V_{C3H8} \times$ (mol H_2O / mol C_3H_8)

> $V_{H2O} = (2.6\ L) \times (4\ mol\ /\ 1\ mol)$

> $V_{H2O} = 10.4\ L$

continued...

24. C is correct.

Use the mnemonic OIL RIG: <u>O</u>xidation <u>I</u>s <u>L</u>oss, <u>R</u>eduction <u>I</u>s <u>G</u>ain (of electrons).

Oxidation is the loss of electrons, while reduction is the gain of electrons.

An oxidizing agent undergoes reduction, while a reducing agent undergoes oxidation.

The oxidation number of Hg decreases (i.e., reduction) from +2 as a reactant ($HgCl_2$) to +1 as a product (Hg_2Cl_2). This means that Hg gained one electron.

25. C is correct.

Balanced equation (synthesis):

$$4\ P\ (s) + 5\ O_2\ (g) \rightarrow 2\ P_2O_5\ (s)$$

26. E is correct.

27. A is correct.

Formula mass is a synonym for molecular mass/molecular weight (MW).

Start by calculating the mass of the formula unit (CH_2O)

$$CH_2O = (12.01\ \text{g/mol} + 2.02\ \text{g/mol} + 16.00\ \text{g/mol}) = 30.03\ \text{g/mol}$$

Divide the molar mass with the formula unit mass:

$$180\ \text{g/mol} / 30.03\ \text{g/mol} = 5.994$$

Round to the closest whole number: 6

Multiply the formula unit by 6:

$$(CH_2O)_6 = C_6H_{12}O_6$$

28. E is correct.

Calculate the moles of LiI present. The molecular weight of LiI is 133.85 g/mol.

Moles of LiI:

$$6.45\ \text{g} / 133.85\ \text{g/mole} = 0.0482\ \text{moles}$$

Each mole of LiI contains 6.02×10^{23} molecules.

Number of molecules:

$$0.0482 \times 6.02 \times 10^{23} = 2.90 \times 10^{22}\ \text{molecules}$$

1 molecule is equal to 1 formula unit.

29. E is correct.

A combustion engine creates various nitrogen oxide compounds, often referred to as NOx gases because each gas would have a different value of x in its formula.

30. C is correct.

Balanced equation (synthesis):

$$4 \text{ P } (s) + 3 \text{ O}_2 (g) \rightarrow 2 \text{ P}_2\text{O}_3 (s)$$

31. A is correct.

Use the mnemonic OIL RIG: <u>O</u>xidation <u>I</u>s <u>L</u>oss, <u>R</u>eduction <u>I</u>s <u>G</u>ain (of electrons).

Oxidation is the loss of electrons, while reduction is the gain of electrons.

An oxidizing agent undergoes reduction, while a reducing agent undergoes oxidation.

Because the forward reaction is spontaneous, the reverse reaction is not spontaneous.

Sn cannot reduce Mg^{2+}; therefore, Sn is the weakest reducing agent.

32. C is correct.

Balancing Redox Equations

From the balanced equation, the coefficient for the proton can be determined. Balancing a redox equation requires balancing the atoms, but the charges must also be balanced.

The equations must be separated into two different half-reactions. One of the half-reactions addresses the oxidizing component, and the other addresses the reducing component.

Magnesium is being oxidized; therefore, the unbalanced oxidation half-reaction will be:

$$\text{Mg } (s) \rightarrow \text{Mg}^{2+} (aq)$$

Furthermore, the nitrogen is being reduced; therefore, the unbalanced reduction half-reaction is:

$$\text{NO}_3^{-} (aq) \rightarrow \text{NO}_2 (aq)$$

At this stage, each half-reaction needs to be balanced for each atom, and the net electric charge on each side of the equations must be balanced.

Order of operations for balancing half-reactions:

1) Balance atoms except for oxygen and hydrogen.

2) Balance the oxygen atom count by adding water.

3) Balance the hydrogen atom count by adding protons.

a) If in basic solution, add equal amounts of hydroxide to each side to cancel protons.

4) Balance the electric charge by adding electrons.

5) If necessary, multiply the coefficients of one half-reaction equation by a factor that cancels the electron count when both equations are combined.

6) Cancel any ions or molecules that appear on both sides of the overall equation.

After determining the balanced overall redox reaction, stoichiometry indicates the moles of protons involved.

For magnesium:

$$Mg\ (s) \rightarrow Mg^{2+}\ (aq)$$

The magnesium is already balanced with a coefficient of 1. There are no hydrogen or oxygen atoms present in the equation. The magnesium cation has a +2 charge, so 2 moles of electrons should be added to the right side to balance the charge.

The balanced half-reaction for oxidation:

$$Mg\ (s) \rightarrow Mg^{2+}\ (aq) + 2\ e^-$$

For nitrogen:

$$NO_3^-\ (aq) \rightarrow NO_2\ (aq)$$

The equation is already balanced for nitrogen because one nitrogen atom appears on both sides of the reaction. The nitrate reactant has three oxygen atoms, while the nitrite product has two oxygen atoms.

To balance the oxygen, one mole of water should be added to the right side:

$$NO_3^-\ (aq) \rightarrow NO_2\ (aq) + H_2O$$

Adding water to the right side of the equation introduces hydrogen atoms to that side. Therefore, the hydrogen atom count needs to be balanced. Water possesses two hydrogen atoms; therefore, two protons need to be added to the left side of the reaction:

$$NO_3^-\ (aq) + 2\ H^+ \rightarrow NO_2\ (aq) + H_2O$$

The reaction occurs in acidic conditions. If the reaction were basic, OH^- would need to be added to both sides to cancel the protons.

Next, the net charge must be balanced. The left side has a net charge of +1 (+2 from the protons and –1 from the electron), while the right side is neutral. Therefore, one electron should be added to the left side:

$$NO_3^-\ (aq) + 2\ H^+ + e^- \rightarrow NO_2\ (aq) + H_2O$$

continued...

When half-reactions are recombined, the electrons in the overall reaction must cancel.

The reduction half-reaction contributes one electron to the left side of the overall equation, while the oxidation half-reaction contributes two electrons to the product side.

Therefore, the coefficients of the reduction half-reaction should be doubled:

$$2 \times [NO_3^- \, (aq) + 2 \, H^+ + e^- \rightarrow NO_2 \, (aq) + H_2O]$$
$$= 2 \, NO_3^- \, (aq) + \mathbf{4 \, H^+} + 2 \, e^- \rightarrow 2 \, NO_2 \, (aq) + 2 \, H_2O$$

The answer is 4 at this step.

Combining both half-reactions gives:

$$Mg \, (s) + 2 \, NO_3^- \, (aq) + \mathbf{4 \, H^+} + 2 \, e^- \rightarrow Mg^{2+} \, (aq) + 2 \, e^- + 2 \, NO_2 \, (aq) + 2 \, H_2O$$

Cancel the electronsfrom both sides of the reaction in the following balanced net equation:

$$Mg \, (s) + 2 \, NO_3^- \, (aq) + \mathbf{4 \, H^+} \rightarrow Mg^{2+} \, (aq) + 2 \, NO_2 \, (aq) + 2 \, H_2O$$

This equation is now fully balanced for mass, oxygen, hydrogen and electric charge.

33. E is correct.

To obtain moles, divide sample mass by the molecular mass.

Because the options have the same mass, a comparison of molecular mass provides the answer without performing the actual calculation.

The molecule with the smallest molecular mass has the greatest number of moles.

34. A is correct.

The law of constant composition states that *all samples of a given chemical compound have the same chemical composition by mass.* This is true for compounds, including ethyl alcohol.

This law is the *law of definite proportions* or *Proust's Law* because this observation was first made by the French chemist Joseph Proust.

35. A is correct.

The coefficients in balanced reactions refer to moles (or molecules) but not grams.

1 mole of N_2 gas reacts with 3 moles of H_2 gas to produce 2 moles of NH_3 gas.

36. C is correct.

Calculate the number of moles of Al_2O_3:

$$2\ Al\ /\ 1\ Al_2O_3 = 0.2\ \text{moles}\ Al\ /\ x\ \text{moles}\ Al_2O_3$$

$$x\ \text{moles}\ Al_2O_3 = 0.2\ \text{moles}\ Al \times 1\ Al_2O_3\ /\ 2\ Al$$

$$x = 0.1\ \text{moles}\ Al_2O_3$$

Therefore:

$$(0.1\ \text{mole}\ Al_2O_3){\cdot}(102\ \text{g/mole}\ Al_2O_3) = 10.2\ \text{g}\ Al_2O_3$$

37. D is correct.

Because Na is a metal, Na is more electropositive than H.

Therefore, the positive charge is on Na (i.e., +1), which means that the charge on H is –1.

38. A is correct.

$PbO + C \rightarrow Pb + CO$ is a single replacement reaction because only one element is being transferred from one reactant to another.

However, it is a redox reaction because the lead (Pb^{2+}), which dissociates, is reduced (gains electrons), the oxygen (O^{2-}) is oxidized (loses electrons).

Additionally, one of the products is a single element (Pb), which is an indication that it is a redox reaction.

C: the reaction is double-replacement and not combustion because the combustion reactions must have hydrocarbon and O_2 as reactants.

Double-replacement reactions often result in a solid, water, and gas formation.

39. B is correct.

The sum of the oxidation numbers in any neutral molecule must be zero.

The oxidation number of each H is +1, and the oxidation number of each O is –2.

If x denotes the oxidation number of S in H_2SO_4, then:

$$2(+1) + x + 4(-2) = 0$$

$$+2 + x + -8 = 0$$

$$x = +6$$

In H_2SO_4, the oxidation sate of sulfur is +6.

40. A is correct.

To calculate the number of molecules, calculate the moles of gas using the ideal gas equation:

$$PV = nRT$$

Rearrange to isolate the number of moles:

$$n = PV / RT$$

Because the gas constant R is in L·atm K^{-1} mol^{-1}, the pressure has to be converted into atm:

$$780 \text{ torr} \times (1 \text{ atm} / 760 \text{ torr}) = 1.026 \text{ atm}$$

Convert the volume to liters:

$$500 \text{ mL} \times 0.001 \text{ L/mL} = 0.5 \text{ L}$$

Substitute into the ideal gas equation:

$$n = PV / RT$$

$$n = 1.026 \text{ atm} \times 0.5 \text{ L} / (0.08206 \text{ L·atm K}^{-1} \text{mol}^{-1} \times 320 \text{ K})$$

$$n = 0.513 \text{ L·atm} / (26.259 \text{ L·atm mol}^{-1})$$

$$n = 0.0195 \text{ mol}$$

Notes for active learning

Practice Set 3: Questions 41–60

41. B is correct.

Use the mnemonic OIL RIG: <u>O</u>xidation <u>I</u>s <u>L</u>oss, <u>R</u>eduction <u>I</u>s <u>G</u>ain (of electrons).

In a redox reaction, the oxidizing and reducing agents are reactants, not products.

The oxidizing agent is the species that gets reduced (i.e., gains electrons) and causes oxidation of the other species.

The reducing agent is the species oxidized (i.e., loses electrons) and causes a reduction of the other species.

42. E is correct.

Balancing a redox equation in acidic solution by the half-reaction method involves each step described.

43. D is correct.

Formula mass is a synonym for molecular mass/molecular weight (MW).

$$\text{MW of } C_6H_{12}O_6 = (6 \times \text{atomic mass C}) + (12 \times \text{atomic mass H}) + (6 \times \text{atomic mass O})$$

$$\text{MW of } C_6H_{12}O_6 = (6 \times 12.01 \text{ g/mole}) + (12 \times 1.01 \text{ g/mole}) + (6 \times 16.00 \text{ g/mole})$$

$$\text{MW of } C_6H_{12}O_6 = (72.06 \text{ g/mole}) + (12.12 \text{ g/mole}) + (96.00 \text{ g/mole})$$

$$\text{MW of } C_6H_{12}O_6 = 180.18 \text{ g/mole}$$

44. E is correct.

The molecular mass of oxygen is 16 g/mole.

The atomic mass unit (amu) or dalton (Da) is the standard unit for indicating mass on an atomic or molecular scale (atomic mass). One amu is approximately the mass of one nucleon (either a single proton or neutron) and is numerically equivalent to 1 g/mol.

45. D is correct.

Balanced equation (combustion):

$$2\,C_2H_6 + 7\,O_2 \rightarrow 4\,CO_2 + 6\,H_2O$$

46. E is correct.

Use the mnemonic OIL RIG: <u>O</u>xidation <u>I</u>s <u>L</u>oss, <u>R</u>eduction <u>I</u>s <u>G</u>ain (of electrons).

Oxidation is the loss of electrons, while reduction is the gain of electrons.

An oxidizing agent undergoes reduction, while a reducing agent undergoes oxidation.

Evaluate the oxidation number in each atom.

S changes from +6 as a reactant to –2 as a product, reducing it because the oxidation number decreases.

47. C is correct.

Balanced equation (double replacement):

$$C_2H_5OH \ (g) + 3 \ O_2 \ (g) \rightarrow 2 \ CO_2 \ (g) + 3 \ H_2O \ (g)$$

48. E is correct.

Note: 1 amu (atomic mass unit) = 1 g/mole.

Moles of C = mass of C / atomic mass of C

Moles of C = (4.50 g) / (12.01 g/mole)

Moles of C = 0.375 mole

49. C is correct.

According to the law of mass conservation, there should be equal amounts of carbon in the product (CO_2) and reactant (the hydrocarbon sample).

Start by calculating the amount of carbon in the product (CO_2).

First, calculate the mass % of carbon in CO_2:

Molecular mass of CO_2 = atomic mass of carbon + (2 × atomic mass of oxygen)

Molecular mass of CO_2 = 12.01 g/mole + (2 × 16.00 g/mole)

Molecular mass of CO_2 = 44.01 g/mole

Mass % of carbon in CO_2 = (mass of carbon / molecular mass of CO_2) × 100%

Mass % of carbon in CO_2 = (12.01 g/mole / 44.01 g/mole) × 100%

Mass % of carbon in CO_2 = 27.3%

continued...

Calculate the mass of carbon in the CO_2:

Mass of carbon = mass of CO_2 × mass % of carbon in CO_2

Mass of carbon = 8.98 g × 27.3%

Mass of carbon = 2.45 g

The mass of carbon in the starting reactant (the hydrocarbon sample) should also be 2.45 g.

Mass % carbon in hydrocarbon = (mass carbon in sample / total mass sample) × 100%

Mass % of carbon in hydrocarbon = (2.45 g / 6.84 g) × 100%

Mass % of carbon in hydrocarbon = 35.8%

50. E is correct.

The number of each atom on the reactants' side must be the same and equal in number to the number of atoms on the product side of the reaction.

51. D is correct.

Using ½ reactions, the balanced reaction is:

$$6\ (Fe^{2+} \rightarrow Fe^{3} + 1\ e^-)$$

$$\frac{Cr_2O_7{}^{2-} + 14\ H^+ + 6\ e^- \rightarrow 2\ Cr^{3+} + 7\ H_2O}{Cr_2O_7{}^{2-} + 14\ H^+ + 6\ Fe^{2+} \rightarrow 2\ Cr^{3+} + 6\ Fe^{3+} + 7\ H_2O}$$

The sum of coefficients in the balanced reaction is 36.

52. E is correct.

Balanced reaction:

$$4\ RuS\ (s) + 9\ O_2 + 4\ H_2O \rightarrow 2\ Ru_2O_3\ (s) + 4\ H_2SO_4$$

Calculate the moles of H_2SO_4:

49 g × 1 mol / 98 g = 0.5 mole H_2SO_4

If 0.5 mole H_2SO_4 is produced, set a ratio for O_2 consumed.

9 O_2 / 4 H_2SO_4 = x moles O_2 / 0.5 mole H_2SO_4

x moles O_2 = 9 O_2 / 4 H_2SO_4 × 0.5 mole H_2SO_4

x = 1.125 moles O_2 × 22.4 liters / 1 mole

x = 25.2 liters of O_2

53. C is correct.

Identify the limiting reactant:

Al is the limiting reactant because 2 atoms of Al combine with 1 molecule of ferric oxide to produce the products.

If the reaction starts with equal numbers of moles of each reactant, Al is depleted.

Use moles of Al and coefficients to calculate moles of other products/reactants.

$$\text{Coefficient of Al} = \text{coefficient of Fe}$$

$$\text{Moles of iron produced} = 0.20 \text{ moles.}$$

54. B is correct.

$$S_2 \qquad O_8$$

$$2x \qquad (8 \times -2)$$

The sum of charges on the molecule is -2:

$$2x + (8 \times -2) = -2$$

$$2x + (-16) = -2$$

$$2x = +14$$

$$\text{oxidation number of S} = +7$$

55. D is correct.

Na is in group IA, so its oxidation number is $+1$.

Oxygen is usually -2, with some exceptions, such as peroxide (H_2O_2), where it is -1.

Cr is a transition metal and could have more than one possible oxidation number.

To determine Cr's oxidation number, use the known oxidation numbers:

$$2(+1) + Cr + 4(-2) = 0$$

$$2 + Cr - 8 = 0$$

$$Cr = 6$$

56. B is correct.

The net charge of $H_2SO_4 = 0$

Hydrogen has a common oxidation number of $+1$, whereas each oxygen atom has an oxidation number of -2.

$$H_2 \qquad\qquad S \qquad\qquad O_4$$
$$(2 \times 1) \qquad\qquad x \qquad\qquad (4 \times -2)$$

$$2 + x + -8 = 0$$

$$x = +6$$

In H_2SO_4, the oxidation of sulfur is $+6$.

57. C is correct.

Use the mnemonic OIL RIG: <u>O</u>xidation <u>I</u>s <u>L</u>oss, <u>R</u>eduction <u>I</u>s <u>G</u>ain (of electrons).

Oxidation is the loss of electrons, while reduction is the gain of electrons.

An oxidizing agent undergoes reduction, while a reducing agent undergoes oxidation.

The oxidation number for Sn is $+2$ as a reactant and $+4$ as a product; an increase in oxidation number means that Sn^{2+} is oxidized.

58. D is correct.

Na_2Cl_2 and $2\ NaCl$ are not the same.

Na_2Cl_2 is a molecule of 2 Na and 2 Cl.

NaCl is a compound that consists of 1 Na and 1 Cl.

59. E is correct.

I: represents the volume of 1 mol of an ideal gas at STP

II: mass of 1 mol of PH_3

III: number of molecules in 1 mol of PH_3

60. D is correct.

Use the mnemonic OIL RIG: <u>O</u>xidation <u>I</u>s <u>L</u>oss, <u>R</u>eduction <u>I</u>s <u>G</u>ain (of electrons).

Oxidation is the loss of electrons, while reduction is the gain of electrons.

An oxidizing agent undergoes reduction, while a reducing agent undergoes oxidation.

Co^{2+} is the starting reactant because it has to lose electrons and produce an ion with a higher oxidation number.

Practice Set 4: Questions 61–80

61. E is correct.

Use the mnemonic OIL RIG: <u>O</u>xidation <u>I</u>s <u>L</u>oss, <u>R</u>eduction <u>I</u>s <u>G</u>ain (of electrons).

In a redox reaction, the oxidizing and reducing agents are reactants, not products.

The oxidizing agent is the reduced species (i.e., gains electrons).

The reducing agent is the oxidized species (i.e., loses electrons).

62. D is correct.

Calculations:

$$\% \text{ mass Cl} = (\text{molecular mass Cl}) / (\text{molecular mass of compound}) \times 100\%$$

$$\% \text{ mass Cl} = 4(35.5 \text{ g/mol}) / [12 \text{ g/mol} + 4(35.5 \text{ g/mol})] \times 100\%$$

$$\% \text{ mass Cl} = (142 \text{ g/mol}) / (154 \text{ g/mol}) \times 100\%$$

$$\% \text{ mass Cl} = 0.922 \times 100\% = 92\%$$

63. E is correct.

I: ionic charge of reactants equals the ionic charge of products, which means that no electrons can "disappear" from the reaction; they can only be transferred from one atom to another.

Therefore, the total charge will be the same.

II: atoms of each reactant must equal atoms of product, which is true for chemical reactions because atoms in chemical reactions cannot be created or destroyed.

III: another atom must lose any electrons gained by one atom.

64. B is correct.

Determine the molecular weight of each compound.

Note, unlike % by mass, there is no division.

$$Cl_2C_2H_4 = 2(35.5 \text{ g/mol}) + 2(12 \text{ g/mol}) + 4(1 \text{ g/mol})$$

$$Cl_2C_2H_4 = 71 \text{ g/mol} + 24 \text{ g/mol} + 4 \text{ g/mol}$$

$$Cl_2C_2H_4 = 99 \text{ g/mol}$$

65. E is correct.

Formula mass is a synonym for molecular mass/molecular weight (MW).

Start by calculating the mass of the formula unit (CH):

$$CH = (12.01 \text{ g/mol} + 1.01 \text{ g/mol}) = 13.02 \text{ g/mol}$$

Divide the molar mass by the formula unit mass:

$$78 \text{ g/mol} / 13.02 \text{ g/mol} = 5.99$$

Round to the closest whole number: 6

Multiply the formula unit by 6:

$$(CH)_6 = C_6H_6$$

66. D is correct.

There is 1 oxygen atom from GaO plus 6 oxygen atoms from $(NO_3)_2$, so the total is 7.

67. B is correct.

Ethanol (C_2H_5OH) undergoes combination with oxygen to produce carbon dioxide and water.

68. C is correct.

Balanced reaction (combustion): $2\ C_3H_7OH + 9\ O_2 \rightarrow 6\ CO_2 + 8\ H_2O$

69. C is correct.

Balanced equation (double replacement):

$$Co_2O_3\ (s) + 3\ CO\ (g) \rightarrow 2\ Co\ (s) + 3\ CO_2\ (g)$$

70. B is correct.

The molar volume (V_m) is the volume occupied by one mole of a substance (i.e., element or compound) at a given temperature and pressure.

71. A is correct.

Use the mnemonic OIL RIG: <u>O</u>xidation <u>I</u>s <u>L</u>oss, <u>R</u>eduction <u>I</u>s <u>G</u>ain (of electrons).

Oxidation is the loss of electrons, while reduction is the gain of electrons.

An oxidizing agent undergoes reduction, while a reducing agent undergoes oxidation.

Because the forward reaction is spontaneous, the reverse reaction is not spontaneous.

Mg^{2+} cannot oxidize Sn; therefore, Mg^{2+} is the weakest oxidizing agent.

72. D is correct.

H_2 is the limiting reactant.

24 moles of H_2 should produce: $24 \times (2 / 3) = 16$ moles of NH_3

If only 13.5 moles are produced, the yield:

13.5 moles / 16 moles $\times$ 100% = 84%

73. B is correct.

Balanced reaction:

$$H_2 + \tfrac{1}{2} O_2 \rightarrow H_2O$$

Multiply the equations by 2 to remove the fraction:

$$2 H_2 + O_2 \rightarrow 2 H_2O$$

Find the limiting reactant by calculating the moles of each reactant:

Moles of hydrogen = mass of hydrogen / (2 $\times$ atomic mass of hydrogen)

Moles of hydrogen = 25 g / (2 $\times$ 1.01 g/mole)

Moles of hydrogen = 12.38 moles

Moles of oxygen = mass of oxygen / (2 $\times$ atomic mass of oxygen)

Moles of oxygen = 225 g / (2 $\times$ 16.00 g/mole)

Moles of oxygen = 7.03 moles

Divide the number of moles of each reactant by its coefficient.

The reactant with a smaller number of moles is the limiting reactant.

Hydrogen = 12.38 moles / 2

Hydrogen = 6.19 moles

Oxygen = 7.03 moles / 1

Oxygen = 7.03 moles

Because hydrogen has a smaller number of moles after the division, hydrogen is the limiting reactant, and hydrogen will be depleted in the reaction.

$$2 H_2 + O_2 \rightarrow 2 H_2O$$

Hydrogen and water have coefficients of 2; therefore, these have the same number of moles.

There are 12.38 moles of hydrogen, which means this reaction produces 12.38 moles of water.

continued...

Molecular mass of H_2O = (2 × molecular mass of hydrogen) + atomic mass of oxygen

Molecular mass of H_2O = (2 × 1.01 g/mole) + 16.00 g/mole

Molecular mass of H_2O = 18.02 g/mole

Mass of H_2O = moles of H_2O × molecular mass of H_2O

Mass of H_2O = 12.38 moles × 18.02 g/mole

Mass of H_2O = 223 g

74. C is correct.

Li	Cl	O_2
+1	x	(2 × –2)

The sum of charges in a neutral molecule is zero:

$1 + x + (2 \times -2) = 0$

$1 + x + (-4) = 0$

$x = -1 + 4$

oxidation number of Cl = +3

75. D is correct.

Using half-reactions, the balanced reaction is:

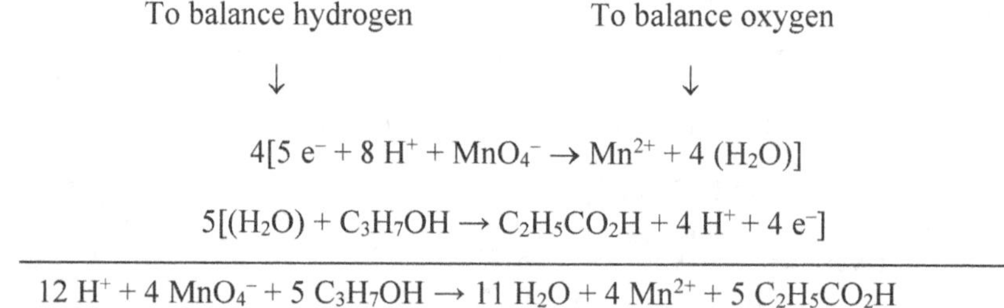

$$\text{To balance hydrogen} \qquad \text{To balance oxygen}$$

$$\downarrow \qquad\qquad \downarrow$$

$$4[5\ e^- + 8\ H^+ + MnO_4^- \rightarrow Mn^{2+} + 4\ (H_2O)]$$

$$5[(H_2O) + C_3H_7OH \rightarrow C_2H_5CO_2H + 4\ H^+ + 4\ e^-]$$

$$12\ H^+ + 4\ MnO_4^- + 5\ C_3H_7OH \rightarrow 11\ H_2O + 4\ Mn^{2+} + 5\ C_2H_5CO_2H$$

Multiply by a common multiple of 4 and 5:

The sum of the product's coefficients: $11 + 4 + 5 = 20$.

76. A is correct.

PbO (s) + C (s) → Pb (s) + CO (g) is a single-replacement reaction because only one element (i.e., oxygen) is being transferred from one reactant to another.

77. E is correct.

The reactants in a chemical reaction are on the left side of the reaction arrow; the products are on the right side.

In this reaction, the reactants are $C_6H_{12}O_6$, H_2O and O_2.

There is a distinction between the terms *reactant* and *reagent.*

A reactant is a substance consumed in the course of a chemical reaction.

A reagent is a substance (e.g., solvent) added to a system to cause a chemical reaction.

78. C is correct.

Determine the number of moles of He:

$$4 \text{ g} \div 4.0 \text{ g/mol} = 1 \text{ mole}$$

Each mole of He has 2 electrons since He has atomic number 2 (# electrons = # protons for neutral atoms).

Therefore, 1 mole of He × 2 electrons / mole = 2 mole of electrons

79. D is correct.

In this molecule, Br has the oxidation number of –1 because it is a halogen and the gaining of one electron results in a complete octet for bromine.

The sum of charges in a neutral molecule = 0

0 = (oxidation state of Fe) + (3 × oxidation state of Br)

0 = (oxidation state of Fe) + (3 × –1)

0 = oxidation state of Fe – 3

oxidation state of Fe = +3

80. B is correct.

A mole is a unit of measurement used to express amounts of a chemical substance.

The number of molecules in a mole is 6.02×10^{23}, which is Avogadro's number.

However, Avogadro's number relates to the number of molecules, not the amount of substance.

Molar mass refers to the mass per mole of a substance.

Formula mass is a term that is sometimes used to mean molecular mass or molecular weight, and it refers to the mass of a specific molecule.

Notes for active learning

===

Practice Set 5: Questions 81–100

===

81. B is correct.

Use the mnemonic OIL RIG: Oxidation Is Loss, Reduction Is Gain (of electrons).

To determine which species undergoes oxidation, determine the oxidation number of species in the reaction.

Reactants: $CuBr$

Br ion is -1, which means that Cu is $+1$.

Products: $Cu = 0$ (it's an element)

$Br_2 = 0$ (it's an element)

Br is the species that undergoes oxidation. It loses electrons and has its oxidation number increase.

82. C is correct.

Use the mnemonic OIL RIG: Oxidation Is Loss, Reduction Is Gain (of electrons).

Oxidation is the loss of electrons, while reduction is the gain of electrons.

An oxidizing agent undergoes reduction, while a reducing agent undergoes oxidation.

Each formula unit of $CuBr_2$ is converted into Cu.

The Br ion does not undergo any changes (i.e., neither oxidation nor reduction); it just moves to Zn to form another compound.

In $CuBr_2$, the oxidation number of Cu is $+2$.

Therefore, when it is reduced, it gains 2 electrons to form Cu.

83. B is correct.

Calculate the molecular mass of CO_2:

Molecular mass of CO_2 = atomic mass of C + ($2 \times$ atomic mass of O)

Molecular mass of CO_2 = 12.01 g/mole + ($2 \times$ 16.00 g/mole)

Molecular mass of CO_2 = 44.01 g/mole

continued...

Calculate the moles of CO_2:

Moles of CO_2 = mass of CO_2 / molecular mass of CO_2

Moles of CO_2 = 168 g / (44.01 g/mole)

Moles of CO_2 = 3.82 moles

One mole of atoms has 6.02×10^{23} atoms (Avogadro's number, or N_A).

Number of CO_2 atoms = moles of CO_2 × Avogadro's number

Number of CO_2 atoms = 3.82 moles × (6.02×10^{23} atoms/mole)

Number of CO_2 atoms = 2.30×10^{24} atoms

84. E is correct.

For calculations, assume 100 g of the compound: 64 g of Ag, 8 g of N and 28 g of O.

Find the number of moles of each of these elements:

Ag: (64 g) / (108 g/mol) = 0.6 mol

N: (8 g) / (14 g/ mol) = 0.6 mol

O: (28 g) / (16 g/mol) = 1.8 mol

Reduce to the smallest coefficients by dividing by 0.6: $AgNO_3$

85. D is correct.

To solve for the density of a gas, use a modification of the ideal gas law.

Ideal gas law:

$PV = nRT$

where n is the number of moles, equal to mass/molecular weight (MW)

Substitute that expression into the ideal gas equation:

$PV = $ (mass / MW) × RT

density (ρ) = mass / volume

Rearrange the equation to get mass / volume on one side:

mass / volume = (P × MW) / RT

density (ρ) = (P × MW) / RT

continued...

Now rearrange the equation to solve for molecular weight (MW):

$$MW = \rho RT / P$$

Here, the gas is in STP condition, with T of 0 °C = 273.15 K and P of 1 atm.

Calculate MW:

$$MW = 1.34 \text{ g/L} \times 0.0821 \text{ L atm K mol}^{-1} \times 273.15 \text{ K} / 1 \text{ atm}$$

$$MW = 30.1 \text{ g}$$

86. C is correct.

Balanced equation (double replacement):

$$C_3H_8\,(g) + 5\ O_2\,(g) \rightarrow 3\ CO_2\,(g) + 4\ H_2O\,(g)$$

87. D is correct. To find what the reaction yields, look at the right side of the equation.

There are 6 CO_2 atoms: $6 \times 2 = 12$

There are 12 H_2O atoms: $12 \times 1 = 12$

$12 + 12 = 24$ atoms of oxygen.

88. C is correct.

Balanced equation (single replacement):

$$2\ Al_2O_3\,(s) + 6\ Cl_2\,(g) \rightarrow 4\ AlCl_3\,(aq) + 3\ O_2\,(g)$$

89. C is correct.

The given half-reaction equation for the reduction of chromium is:

$$Cr_2O_7^{2-}\,(aq) \rightarrow Cr^{3+}\,(aq)$$

where the oxidation state of oxygen in $Cr_2O_7^{2-}$ is –2, and the oxidation state of chromium is +6

2(charge of metal) + 7(charge of oxygen) = net charge of the molecule or ion

$$2(Cr^{n+}) + 7(-2) = -2$$

$$2(Cr^{n+}) - 14 = -2$$

$$2(Cr^{n+}) = 12$$

$$Cr^{n+} = +6$$

continued...

To balance this reduction half-reaction:

1) Balance the metal count.

2) Balance oxygen count.

3) Balance hydrogen count.

4) Add hydroxide if the reaction is done in base.

5) Balance the charge.

6) Optional step: multiply the coefficients by a factor to cancel electrons, if the reaction is to be combined with the oxidation half-reaction

Applying these operations to the half-reaction: $Cr_2O_7^{2-} \rightarrow Cr^{3+}$

1) Balance the metal count on both sides of the equation by adding the appropriate coefficients:

$$Cr_2O_7^{2-} \rightarrow 2\ Cr^{3+}$$

2) Balance the oxygen count by adding water:

$$Cr_2O_7^{2-} \rightarrow 2\ Cr^{3+} + 7\ H_2O$$

3) Balance the hydrogen count by adding protons:

$$14\ H^+ + Cr_2O_7^{2-} \rightarrow 2\ Cr^{3+} + 7\ H_2O$$

Skip step 4.

5) Balance the charge by adding electrons (there should be a net charge of +6 on both sides):

$$14\ H^+ + Cr_2O_7^{2-} + 6\ e^- \rightarrow 2\ Cr^{3+} + 7\ H_2O$$

90. B is correct.

Na	H	C	O_3
+1	+1	x	(3×-2)

The sum of charges in a neutral molecule $= 0$

$$1 + 1 + x + (3 \times -2) = 0$$

$$1 + 1 + x + (-6) = 0$$

$$x = -1 - 1 + 6$$

$$x = +4$$

oxidation number of C $= +4$

91. D is correct.

Avogadro's law states that the relationship between the masses of the same volume of the same gases (at the same temperature and pressure) corresponds to the relationship between their respective molecular weights. Hence, the relative molecular mass of a gas can be calculated from the mass of a sample of known volume.

The flaw in Dalton's theory was corrected in 1811 by Avogadro, who proposed that equal volumes of any two gases, at equal temperature and pressure, contain equal numbers of molecules (i.e., the mass of a gas's particles does not affect the volume that it occupies).

Avogadro's law allowed him to deduce the diatomic nature of numerous gases by studying the volumes at which they reacted. For example, two liters of hydrogen react with just one liter of oxygen to produce two liters of water vapor (at constant pressure and temperature). This means that a single oxygen molecule (O_2) splits to form two water particles.

Thus, Avogadro offered more accurate estimates of the atomic mass of oxygen and various other elements and distinguished between molecules and atoms.

92. B is correct.

Balanced reaction: $BaCl_2$ (*aq*) + K_2SO_4 (*aq*) $\rightarrow$ $BaSO_4$ and 2 KCl

A double replacement reaction indicates an exchange of cations and anions between the reactants.

Separate the reactants into ions and then exchange the cation and anion pairings.

93. B is correct.

$$\% \text{ mass Cl} = (\text{mass of chlorine in molecule} / \text{molecular mass}) \times 100\%$$

$$\% \text{ mass Cl} = (2 \times 35.45 \text{ g/mol}) / 159.09 \text{ g/mol} \times 100\%$$

$$\% \text{ mass Cl} = 0.446 \times 100\% = 44.6\%$$

94. A is correct.

Use the mnemonic OIL RIG: <u>O</u>xidation <u>Is</u> <u>L</u>oss, <u>R</u>eduction <u>Is</u> <u>G</u>ain (of electrons).

Oxidation is the loss of electrons, while reduction is the gain of electrons.

An oxidizing agent undergoes reduction, while a reducing agent undergoes oxidation.

Because the forward reaction is spontaneous, the reverse reaction is not spontaneous.

I_2 cannot oxidize $FeCl_2$; therefore, I_2 is the weakest oxidizing agent.

NaCl is not the answer because Na and Cl ions do not participate in the redox reaction; each has an identical oxidation number on both sides of the reaction.

95. C is correct.

Balanced reaction:

$$4 \text{ RuS } (s) + 9 \text{ O}_2 + 4 \text{ H}_2\text{O} \rightarrow 2 \text{ Ru}_2\text{O}_3 (s) + 4 \text{ H}_2\text{SO}_4$$

Find moles of RuS:

22.4 liters O_2 at STP = 1 mole O_2

67 g RuS $\times$ 1 mole/133 g = 0.5 mole RuS

Since 4 RuS molecules combine with 9 O_2 molecules, given 1 mole of O_2 and 0.5 mole RuS, O_2 is the limiting reactant.

Ratio for the limiting reactant:

molar ratio: 9 O_2 / 4 RuS = 1 mole O_2 / x mole RuS

4 RuS / 9 O_2 = x mole RuS / 1 mole O_2

x mole RuS / 1 mole O_2 = 4 RuS / 9 O_2

x mole RuS = 4 RuS / 9 O_2 $\times$ 1 mole O_2

$x = (4 / 9)$ mole RuS

Since RuS and H_2SO_4 have the same ratio, 4 / 9 H_2SO_4 is produced.

$(4 / 9)$ mole H_2SO_4 $\times$ 98 g/mole = 44 g

96. A is correct.

Ca	S	O_4
+2	x	(4×-2)

The sum of charges in a neutral molecule is zero:

$2 + x + (4 \times -2) = 0$

$2 + x + (-8) = 0$

$x = -2 + 8$

oxidation number of S = +6

97. C is correct.

Balanced double replacement reaction:

$$C_6H_{12}O_6 (s) + 6 \text{ O}_2 (g) \rightarrow 6 \text{ CO}_2 (g) + 6 \text{ H}_2\text{O} (g)$$

98. B is correct.

$SO_3 + H_2O \rightarrow H_2SO_4$ is a synthesis reaction because two compounds produce one product.

$3\ CuSO_4 + 2\ Al \rightarrow Al_2(SO_4)_3 + 3\ Cu$ is a single-replacement reaction because only one element (i.e., SO_4) is transferred from one reactant to another.

$2\ NaHCO_3 \rightarrow Na_2CO_3 + CO_2 + H_2O$ is a decomposition reaction where a compound is broken down into constituent elements.

$C_3H_8 + 5\ O_2 \rightarrow 3\ CO_2 + 4\ H_2O$ is a combustion reaction because a hydrocarbon (i.e., propane) reacts with oxygen to produce carbon dioxide and water.

99. A is correct.

The coefficients give the stoichiometric number (i.e., relative quantities) of molecules (or atoms) involved in the reaction.

100. D is correct.

Calculate the molecular mass of NO:

Molecular mass of NO = atomic mass of N + atomic mass of O

Molecular mass of NO = N: 14.01 g/mole + O: 16.00 g/mole

Molecular mass of NO = 30.01 g/mole

Calculate moles of NO:

Moles of NO = mass of NO / molecular mass of NO

Moles of NO = 17.0 g / 30.01 g/mole

Moles of NO = 0.57 mole

For STP conditions, 1 mole of gas has a volume of 22.4 L:

Volume of NO = moles of NO × 22.4 L/mole

Volume of NO = 0.57 mole × 22.4 L

Volume of NO = 12.7 L

Notes for active learning

===

Practice Set 6: Questions 101–120

===

101. B is correct.

Use the mnemonic OIL RIG: <u>O</u>xidation <u>I</u>s <u>L</u>oss, <u>R</u>eduction <u>I</u>s <u>G</u>ain (of electrons).

Oxidizing agents undergo reduction (gain electrons) in a redox reaction.

Cl^- has a full octet electron configuration and has a slight tendency to pick up more electrons.

Na^+ has a full octet electron configuration.

Sodium metal is very reactive, and it is a strong reducing agent. As a result, the conjugate ion (Na^+) will be a weak oxidizing agent.

Cl_2 is already in a stable electron configuration, but it can be reduced to Cl^-.

Cl_2 is very reactive and a strong oxidizing agent.

102. B is correct.

Use the mnemonic OIL RIG: <u>O</u>xidation <u>I</u>s <u>L</u>oss, <u>R</u>eduction <u>I</u>s <u>G</u>ain (of electrons).

Oxidation is the loss of electrons, while reduction is the gain of electrons.

An oxidizing agent undergoes reduction, while a reducing agent undergoes oxidation.

If the substance loses electrons (being oxidized), another substance must be gaining electrons (being reduced).

The original substance is referred to as a reducing agent, even though it is being oxidized.

103. A is correct.

A molecular formula expresses the number of atoms of each element in a molecule.

An empirical formula is the simplest formula for a compound.

For example, if the molecular formula is C_6H_{16}, the empirical formula is C_3H_8.

Elemental formula and atomic formula are not valid chemistry terms.

104. E is correct.

Use the mnemonic OIL RIG: <u>O</u>xidation <u>I</u>s <u>L</u>oss, <u>R</u>eduction <u>I</u>s <u>G</u>ain (of electrons).

Oxidation is the loss of electrons, while reduction is the gain of electrons.

An oxidizing agent undergoes reduction, while a reducing agent undergoes oxidation.

Because the forward reaction is spontaneous, the reverse reaction is not spontaneous.

$FeCl_2$ cannot reduce I_2; therefore, $FeCl_2$ is the weakest reducing agent.

NaCl is not the answer because Na and Cl ions do not participate in the redox reaction; each has an identical oxidation number on both sides of the reaction.

105. C is correct.

Use the atomic mass from the periodic table:

$$(C = 6 \times 12 \text{ grams}) + (H = 12 \times 1 \text{ gram}) + (O = 6 \times 16 \text{ grams}) = 180 \text{ g/mole}$$

$$180 \text{ g/mole} \times 3.5 \text{ moles} = 630 \text{ grams}$$

106. B is correct.

The number of oxygens on both sides of the reaction equation must be equal.

On the right side, there are 8 CO_2 molecules; since each CO_2 molecule contains 2 oxygens, multiply $8 \times 2 = 16$.

On the right side are 10 H_2O molecules; $10 \times 1 = 10$.

Add 16 and 10 to get the number of oxygens on the right side: $16 + 10 = 26$.

Since there are 26 oxygens on the right side, there should be 26 oxygens on the left side.

Each O_2 molecule contains 2 oxygens; $26 / 2 = 13$.

Therefore, coefficient 13 is needed to balance the equation.

$$2 \, C_4H_{10} \, (g) + 13 \, O_2 \, (g) \rightarrow 8 \, CO_2 \, (g) + 10 \, H_2O \, (g)$$

107. D is correct.

A mole is a unit of measurement used to express amounts of a chemical substance.

Avogadro's number is the number of constituent particles (i.e., often atoms or molecules) equals approximately 6.022×10^{23}.

Avogadro's number equals the number of atoms in 1 mole (g atomic weight) of an element.

The number of molecules in a mole is 6.02×10^{23}, which is Avogadro's number.

108. A is correct.

The balanced equation is:

$$2 \text{ RuS} + 9/2 \text{ O}_2 + 2 \text{ H}_2\text{O} \rightarrow \text{Ru}_2\text{O}_3 + 2 \text{ H}_2\text{SO}_4$$

To avoid fractional coefficients, multiply coefficients by 2:

$$4 \text{ RuS} + 9 \text{ O}_2 + 4 \text{ H}_2\text{O} \rightarrow 2 \text{ Ru}_2\text{O}_3 + 4 \text{ H}_2\text{SO}_4$$

The sum of the coefficients:

$$4 + 9 + 4 + 2 + 4 = 23$$

109. C is correct.

The % by mass alone is not sufficient because it provides only the empirical formula.

The molecular weight alone is not sufficient because two different compounds might have similar molecular weights.

If molecular mass and % by mass are known, the empirical formula is multiplied by a factor to determine the molecular weight.

110. E is correct.

Balanced equation (synthesis) for reaction II: $4 \text{ Al } (s) + 3 \text{ Br}_2 (l) \rightarrow 2 \text{ Al}_2\text{Br}_3 (s)$

111. B is correct.

First, calculate the molecular mass (MW) of $MgSO_4$:

MW of $MgSO_4$ = atomic mass of Mg + atomic mass of S + (4 × atomic mass of O)

MW of $MgSO_4$ = 24.3 g/mole + 32.1 g/mole + (4 × 16.0 g/mole)

MW of $MgSO_4$ = 120.4 g/mole

Then, use the molecular mass to determine the number of moles:

Moles of $MgSO_4$ = mass of $MgSO_4$ / MW of $MgSO_4$

Moles of $MgSO_4$ = 60.2 g / 120.4 g/mole

Moles of $MgSO_4$ = 0.5 mole

112. A is correct.

Grams of a compound often contain a very large number of atoms.

72.9 g of Mg × (1 mol/24.3 g) = 3 moles Mg

3 moles Mg × (6.02×10^{23} atoms/mol) = 1.81×10^{24} atoms

113. A is correct.

To obtain the percent mass composition of oxygen, calculate the mass of oxygen in the molecule, divide it by the molecular mass and multiply by 100%.

$$\% \text{ mass oxygen} = (\text{mass of oxygen in molecule} / \text{molecular mass}) \times 100\%$$

$$\% \text{ mass oxygen} = [(4 \text{ moles} \times 16.00 \text{ g/mole}) / (303.39 \text{ g/mole})] \times 100\%$$

$$\% \text{ mass oxygen} = 0.211 \times 100\% = 21.1\%$$

114. A is correct.

Balanced reaction:

$$4 \text{ RuS } (s) + 9 \text{ O}_2 + 4 \text{ H}_2\text{O} \rightarrow 2 \text{ Ru}_2\text{O}_3 (s) + 4 \text{ H}_2\text{SO}_4$$

Set up a ratio and solve:

$$4 \text{ RuS} / 2 \text{ Ru}_2\text{O}_3 = 7 \text{ moles RuS} / x \text{ moles Ru}_2\text{O}_3$$

$$x \text{ moles Ru}_2\text{O}_3 = 4 \text{ RuS} \times 7 \text{ moles RuS} / 2 \text{ Ru}_2\text{O}_3$$

$$x \text{ moles Ru}_2\text{O}_3 = 7 \text{ moles RuS} \times 2 \text{ Ru}_2\text{O}_3 / 4 \text{ RuS}$$

$$x = 3.5 \text{ moles}$$

115. C is correct.

Oxidation numbers of S in compounds:

SO_4^{2-}	+6
$S_2O_3^{2-}$	+2
S^{2-}	−2

116. B is correct.

Li is in group IA, and its oxidation number is +1.

Therefore, in Li_2O_2, the oxidation number of oxygen is −1.

Oxygen has an oxidation number of −1 (like in peroxides, H_2O_2).

117. C is correct.

A double-replacement reaction occurs when parts of two ionic compounds (e.g., HBr and KOH) are exchanged to make two new compounds (H_2O and KBr).

$2\,HI \rightarrow H_2 + I_2$ is a decomposition reaction where a compound breaks down into constituent elements.

$SO_2 + H_2O \rightarrow H_2SO_4$ is a synthesis (or composition) reaction when two species combine to form a more complex chemical product.

$CuO + H_2 \rightarrow Cu + H_2O$ is a single-replacement reaction when one element is transferred from one reactant to another (i.e., the oxygen atom leaves the Cu and joins H_2).

118. B is correct.

Density = mass / volume

$$\text{Density} = (5\ \mu g \times 1g\ /1{,}000{,}000\ \mu g) / (25\ \mu L \times 1\ L\ /\ 1{,}000{,}000\ \mu L)$$

$$\text{Density} = (5 \times 10^{-6}\ g) / (2.5 \times 10^{-5}\ L)$$

$$\text{Density} = 0.2\ g/L$$

Perform the above calculation (i.e., density = mass/volume) for each answer choice to determine the lowest density.

119. E is correct.

Balanced double replacement reaction:

$$2\,C_6H_{14}\,(g) + 19\,O_2\,(g) \rightarrow 12\,CO_2\,(g) + 14\,H_2O\,(g)$$

120. A is correct.

A mole is a unit of measurement used to express amounts of a chemical substance.

The number of molecules in a mole is 6.02×10^{23}, which is Avogadro's number.

Notes for active learning

Notes for active learning

Notes for active learning

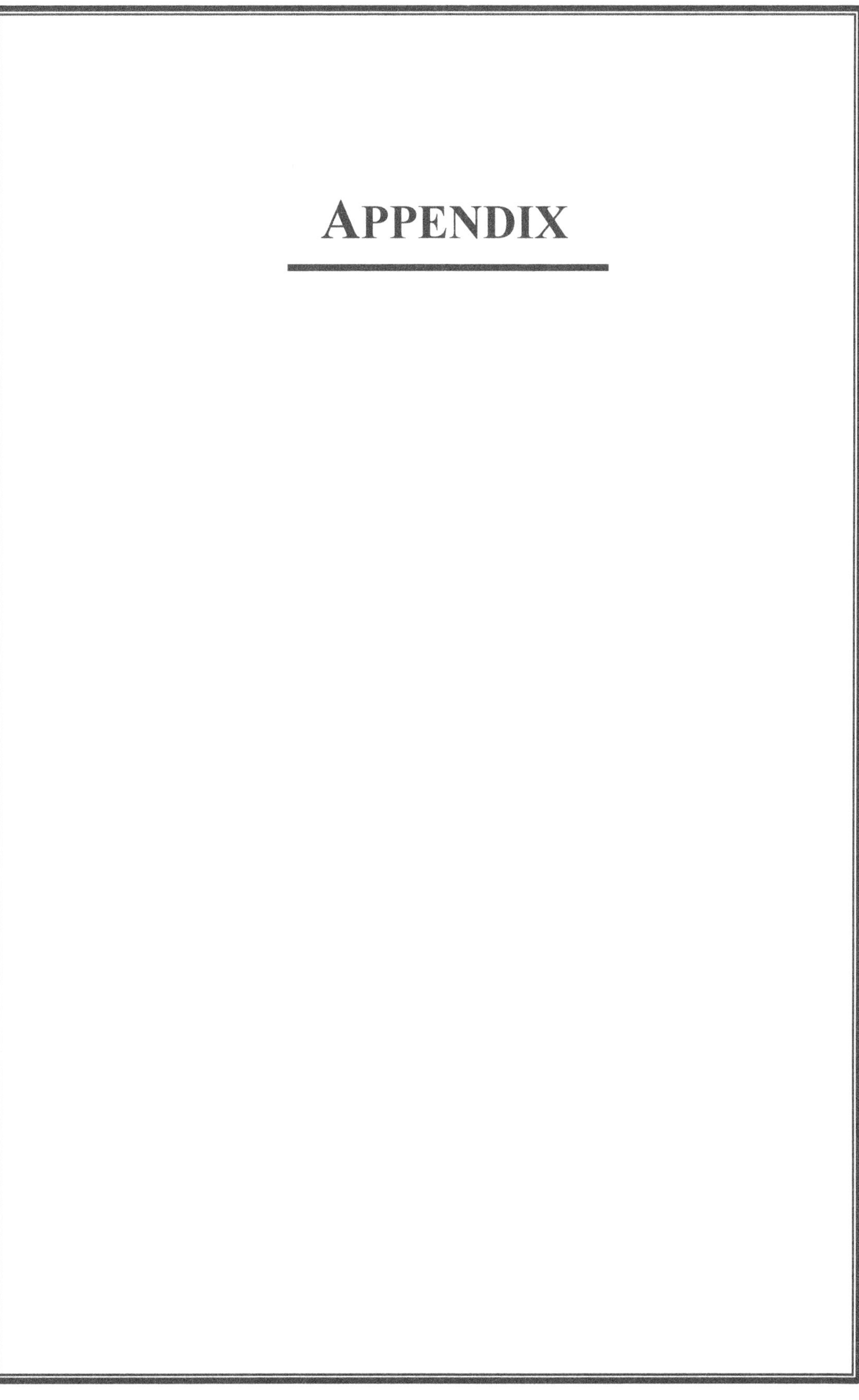

APPENDIX

Periodic Table of the Elements

Key:
atomic number
Symbol
name
abridged standard atomic weight

1	2	3	4	5	6	7	8	9	10	11	12	13	14	15	16	17	18
1 H hydrogen 1.0080 ±0.0002																	**2 He** helium 4.0026 ±0.0001
3 Li lithium 6.94 ±0.06	**4 Be** beryllium 9.0122 ±0.0001											**5 B** boron 10.81 ±0.02	**6 C** carbon 12.011 ±0.002	**7 N** nitrogen 14.007 ±0.001	**8 O** oxygen 15.999 ±0.001	**9 F** fluorine 18.998 ±0.001	**10 Ne** neon 20.180 ±0.001
11 Na sodium 22.990 ±0.001	**12 Mg** magnesium 24.305 ±0.002											**13 Al** aluminium 26.982 ±0.001	**14 Si** silicon 28.085 ±0.001	**15 P** phosphorus 30.974 ±0.001	**16 S** sulfur 32.06 ±0.02	**17 Cl** chlorine 35.45 ±0.01	**18 Ar** argon 39.95 ±0.16
19 K potassium 39.098 ±0.001	**20 Ca** calcium 40.078 ±0.004	**21 Sc** scandium 44.956 ±0.001	**22 Ti** titanium 47.867 ±0.001	**23 V** vanadium 50.942 ±0.001	**24 Cr** chromium 51.996 ±0.001	**25 Mn** manganese 54.938 ±0.001	**26 Fe** iron 55.845 ±0.002	**27 Co** cobalt 58.933 ±0.001	**28 Ni** nickel 58.693 ±0.001	**29 Cu** copper 63.546 ±0.003	**30 Zn** zinc 65.38 ±0.02	**31 Ga** gallium 69.723 ±0.001	**32 Ge** germanium 72.630 ±0.008	**33 As** arsenic 74.922 ±0.001	**34 Se** selenium 78.971 ±0.008	**35 Br** bromine 79.904 ±0.003	**36 Kr** krypton 83.798 ±0.002
37 Rb rubidium 85.468 ±0.001	**38 Sr** strontium 87.62 ±0.01	**39 Y** yttrium 88.906 ±0.001	**40 Zr** zirconium 91.224 ±0.002	**41 Nb** niobium 92.906 ±0.001	**42 Mo** molybdenum 95.95 ±0.01	**43 Tc** technetium [97]	**44 Ru** ruthenium 101.07 ±0.02	**45 Rh** rhodium 102.91 ±0.01	**46 Pd** palladium 106.42 ±0.01	**47 Ag** silver 107.87 ±0.01	**48 Cd** cadmium 112.41 ±0.01	**49 In** indium 114.82 ±0.01	**50 Sn** tin 118.71 ±0.01	**51 Sb** antimony 121.76 ±0.01	**52 Te** tellurium 127.60 ±0.03	**53 I** iodine 126.90 ±0.01	**54 Xe** xenon 131.29 ±0.01
55 Cs caesium 132.91 ±0.01	**56 Ba** barium 137.33 ±0.01	**57-71** lanthanoids	**72 Hf** hafnium 178.49 ±0.01	**73 Ta** tantalum 180.95 ±0.01	**74 W** tungsten 183.84 ±0.01	**75 Re** rhenium 186.21 ±0.01	**76 Os** osmium 190.23 ±0.03	**77 Ir** iridium 192.22 ±0.01	**78 Pt** platinum 195.08 ±0.02	**79 Au** gold 196.97 ±0.01	**80 Hg** mercury 200.59 ±0.01	**81 Tl** thallium 204.38 ±0.01	**82 Pb** lead 207.2 ±1.1	**83 Bi** bismuth 208.98 ±0.01	**84 Po** polonium [209]	**85 At** astatine [210]	**86 Rn** radon [222]
87 Fr francium [223]	**88 Ra** radium [226]	**89-103** actinoids	**104 Rf** rutherfordium [267]	**105 Db** dubnium [268]	**106 Sg** seaborgium [269]	**107 Bh** bohrium [270]	**108 Hs** hassium [269]	**109 Mt** meitnerium [277]	**110 Ds** darmstadtium [281]	**111 Rg** roentgenium [282]	**112 Cn** copernicium [285]	**113 Nh** nihonium [286]	**114 Fl** flerovium [290]	**115 Mc** moscovium [290]	**116 Lv** livermorium [293]	**117 Ts** tennessine [294]	**118 Og** oganesson [294]

Lanthanoids (57–71)

57	58	59	60	61	62	63	64	65	66	67	68	69	70	71
La lanthanum 138.91 ±0.01	**Ce** cerium 140.12 ±0.01	**Pr** praseodymium 140.91 ±0.01	**Nd** neodymium 144.24 ±0.01	**Pm** promethium [145]	**Sm** samarium 150.36 ±0.02	**Eu** europium 151.96 ±0.01	**Gd** gadolinium 157.25 ±0.03	**Tb** terbium 158.93 ±0.01	**Dy** dysprosium 162.50 ±0.01	**Ho** holmium 164.93 ±0.01	**Er** erbium 167.26 ±0.01	**Tm** thulium 168.93 ±0.01	**Yb** ytterbium 173.05 ±0.02	**Lu** lutetium 174.97 ±0.01

Actinoids (89–103)

89	90	91	92	93	94	95	96	97	98	99	100	101	102	103
Ac actinium [227]	**Th** thorium 232.04 ±0.01	**Pa** protactinium 231.04 ±0.01	**U** uranium 238.03 ±0.01	**Np** neptunium [237]	**Pu** plutonium [244]	**Am** americium [243]	**Cm** curium [247]	**Bk** berkelium [247]	**Cf** californium [251]	**Es** einsteinium [252]	**Fm** fermium [257]	**Md** mendelevium [258]	**No** nobelium [259]	**Lr** lawrencium [262]

International Union of Pure and Applied Chemistry (IUPAC), 4 May 2022

Notes for active learning

Common Chemistry Equations

Throughout the test the following symbols have the definitions specified unless otherwise noted.

L, mL	= liter(s), milliliter(s)		mm Hg	= millimeters of mercury
g	= gram(s)		J, kJ	= joule(s), kilojoule(s)
nm	= nanometer(s)		V	= volt(s)
atm	= atmosphere(s)		mol	= mole(s)

ATOMIC STRUCTURE

$$E = h\nu$$

$$c = \lambda\nu$$

E = energy

ν = frequency

λ = wavelength

Planck's constant, $h = 6.626 \times 10^{-34}$ J s

Speed of light, $c = 2.998 \times 10^{8}$ m s^{-1}

Avogadro's number $= 6.022 \times 10^{23}$ mol^{-1}

Electron charge, $e = -1.602 \times 10^{-19}$ coulomb

EQUILIBRIUM

$$K_c = \frac{[C]^c[D]^d}{[A]^a[B]^b}, \text{ where } a\,A + b\,B \rightleftarrows c\,C + d\,D$$

$$K_p = \frac{(P_C)^c(P_D)^d}{(P_A)^a(P_B)^b}$$

$$K_a = \frac{[H^+][A^-]}{[HA]}$$

$$K_b = \frac{[OH^-][HB^+]}{[B]}$$

$$K_w = [H^+][OH^-] = 1.0 \times 10^{-14} \text{ at } 25°C$$
$$= K_a \times K_b$$

$$pH = -\log[H^+], \ pOH = -\log[OH^-]$$

$$14 = pH + pOH$$

$$pH = pK_a + \log\frac{[A^-]}{[HA]}$$

$$pK_a = -\log K_a, \ pK_b = -\log K_b$$

Equilibrium Constants

K_c (molar concentrations)

K_p (gas pressures)

K_a (weak acid)

K_b (weak base)

K_w (water)

KINETICS

$$\ln[A]_t - \ln[A]_0 = -kt$$

$$\frac{1}{[A]_t} - \frac{1}{[A]_0} = kt$$

$$t_{1/2} = \frac{0.693}{k}$$

k = rate constant

t = time

$t_{1/2}$ = half-life

GASES, LIQUIDS, AND SOLUTIONS

$$PV = nRT$$

$$P_A = P_{total} \times X_A, \text{ where } X_A = \frac{\text{moles A}}{\text{total moles}}$$

$$P_{total} = P_A + P_B + P_C + \ldots$$

$$n = \frac{m}{M}$$

$$K = {}^\circ C + 273$$

$$D = \frac{m}{V}$$

$$KE \text{ per molecule} = \frac{1}{2}mv^2$$

Molarity, M = moles of solute per liter of solution

$$A = abc$$

P = pressure
V = volume
T = temperature
n = number of moles
m = mass
M = molar mass
D = density
KE = kinetic energy
v = velocity
A = absorbance
a = molar absorptivity
b = path length
c = concentration

Gas constant, R = 8.314 J mol^{-1} K^{-1}
$\qquad$ = 0.08206 L atm mol^{-1} K^{-1}
$\qquad$ = 62.36 L torr mol^{-1} K^{-1}
1 atm = 760 mm Hg
$\qquad$ = 760 torr
STP = 0.00 °C and 10^5 Pa

THERMOCHEMISTRY/ ELECTROCHEMISTRY

$$q = mc\Delta T$$

$$\Delta S^\circ = \sum S^\circ \text{ products} - \sum S^\circ \text{ reactants}$$

$$\Delta H^\circ = \sum \Delta H_f^\circ \text{ products} - \sum \Delta H_f^\circ \text{ reactants}$$

$$\Delta G^\circ = \sum \Delta G_f^\circ \text{ products} - \sum \Delta G_f^\circ \text{ reactants}$$

$$\Delta G^\circ = \Delta H^\circ - T\Delta S^\circ$$

$$= -RT \ln K$$

$$= -n F E^\circ$$

$$I = \frac{q}{t}$$

q = heat
m = mass
c = specific heat capacity
T = temperature
S° = standard entropy
H° = standard enthalpy
G° = standard free energy
n = number of moles
E° = standard reduction potential
I = current (amperes)
q = charge (coulombs)
t = time (seconds)

Faraday's constant, F = 96,485 coulombs per mole
$\qquad$ of electrons

$$1 \text{ volt} = \frac{1 \text{ joule}}{1 \text{ coulomb}}$$

Annotated Glossary of Chemistry Terms

A

Absolute entropy (of a substance) – the increase in the entropy of a substance as it goes from a perfectly ordered crystalline form at 0 K (where its entropy is zero) to the temperature in question.

Absolute zero – the zero point on the absolute temperature scale; –273.15 °C or 0 K; theoretically, the temperature at which molecular motion ceases (i.e., the system does not emit or absorb energy, atoms at rest).

Absorption spectrum – spectrum associated with the absorption of electromagnetic radiation by atoms (or other species), resulting from transitions from lower to higher energy states.

Accuracy – the degree to which a value is close to the actual value; see *precision*.

Acid – a substance that produces H^+ (aq) ions in an aqueous solution and gives a pH of less than 7.0; strong acids ionize entirely or almost entirely in dilute aqueous solution; weak acids ionize only slightly. It turns litmus red.

Acid dissociation constant – an equilibrium constant for dissociating a weak acid.

Acid rain – rainwater with a pH of less than 5.7; caused by the gases NO_2 (vehicle exhaust fumes) and SO_2 (from burning fossil fuels) dissolving in the rain. It kills fish, wildlife, and trees and destroys buildings and lakes.

Acidic salt – contains an ionizable hydrogen atom; does not necessarily produce acidic solutions.

Actinides – the fifteen chemical elements that are between actinium (89) and lawrencium (103).

Activated complex – a structure formed by a collision between molecules in which new bonds form.

Activation energy – the amount of energy that reactants must absorb in their ground states to reach the transition state needed for a reaction to occur.

Active metal – a metal with low ionization energy that loses electrons readily to form cations.

Activity (of a component of ideal mixture) – a dimensionless quantity whose magnitude is equal to the molar concentration in an ideal solution; equal to partial pressure in an ideal gas mixture; 1 for pure solids or liquids.

Activity series – a listing of metals (and hydrogen) in order of decreasing activity.

Actual yield – the amount of a specified pure product obtained from a given reaction; see *theoretical yield*.

Addition reaction – a reaction in which two atoms or groups of atoms are added to a molecule, one on each side of a double or triple bond.

Adhesive forces – forces of attraction between a liquid and another surface.

Adsorption – the adhesion of a species onto the surfaces of particles.

Aeration – the mixing of air into a liquid or a solid.

Alcohol – a hydrocarbon derivative containing a ~OH group attached to a carbon atom, not in an aromatic ring.

Alkali metals – the elements of Group IA on the periodic table (e.g., Na, K, Rb).

Alkaline battery – a dry cell in which the electrolyte contains KOH.

Alkaline earth metals – group IIA metals on the periodic table; see *earth metals*.

Allomer – a substance that has a different composition from another but the same crystalline structure.

Allotropes – elements with different structures (therefore different forms), such as carbon (e.g., diamond, graphite, and fullerene).

Allotropic modifications (allotropes) – different forms of the same element in the same physical state.

Alloy – a mixture of metals. For example, bronze is an alloy formed from copper and tin.

Alloying – mixing metals with other substances (usually other metals) to modify their properties.

Alpha (α) particle – a helium nucleus; helium ion with 2+ charge; an assembly of two protons and two neutrons.

Amorphous solid – a non-crystalline solid with no well-defined ordered structure.

Ampere – unit of electrical current; one ampere equals one coulomb per second.

Amphiprotism – the ability of a substance to exhibit amphiprotic by accepting donated protons.

Amphoterism – the ability to react with both acids and bases; to act as either an acid or a base.

Amplitude – the maximum distance that medium particles carrying the wave move from their rest position.

Anion – a negative ion; an atom or group of atoms that has gained one or more electrons.

Anode – in a cathode ray tube, the positive electrode (electrode at which oxidation occurs); the positive side of a dry cell battery or a cell.

Antibonding orbital – a molecular orbital higher in energy than any of the atomic orbitals from which it is derived; lends instability to a molecule or ion when populated with electrons; denoted with star (*) superscript.

Artificial transmutation – an artificially induced nuclear reaction caused by the bombardment of a nucleus with subatomic particles or small nuclei.

Associated ions – short-lived species formed by the collision of dissolved ions of opposite charges.

Atmosphere – a unit of pressure; the pressure supports a column of mercury 760 mm high at 0 °C.

Atom – a chemical element in its smallest form; it comprises neutrons and protons within the nucleus and electrons circling the nucleus; it is the smallest part of an element that can exist.

Atomic mass unit (amu) – one-twelfth of the mass of an atom of the carbon-12 isotope; used for stating atomic and formula weights; known as a dalton.

Atomic number – represents an element corresponding with the number of protons within the nucleus; the number of protons in the nucleus of the atom.

Atomic orbital (*AO*) – a region or volume in space where the probability of finding electrons is highest.

Atomic radius – radius of an atom.

Atomic weight – weighted average of the masses of the constituent isotopes of an element; the relative masses of atoms of different elements.

Aufbau (or *building up*) principle – describes the order in which electrons fill orbitals in atoms.

Autoionization – an ionization reaction between identical molecules.

Avogadro's Law – equal volumes of gases contain the same number of molecules at the same temperature and pressure.

Avogadro's number (N_A) – the number (6.022×10^{23}) of atoms, molecules, or particles found in precisely 1 mole of a substance.

B

Background radiation – extraneous to an experiment; usually the low-level natural radiation from cosmic rays and trace radioactive substances present in the environment.

Band – a series of very closely spaced nearly continuous molecular orbitals that belong to the crystal as a whole.

Band of stability – band containing nonradioactive nuclides in a plot of neutrons *vs.* their atomic number.

Band theory of metals – the theory that accounts for the bonding and properties of metallic solids.

Barometer – a device used to measure the pressure in the atmosphere.

Base – a substance that produces $^-$OH (*aq*) ions in an aqueous solution; accepts a proton and has a high pH; strongly soluble bases are soluble in water and are entirely dissociated; weak bases ionize only slightly; a typical example of a base is sodium hydroxide (NaOH). It turns litmus blue.

Basic anhydride – the oxide of a metal that reacts with water to form a base.

Basic salt – a salt containing an ionizable OH group.

Beta (β) particle – an electron emitted from the nucleus when a neutron decays to a proton and an electron.

Binary acid – a binary compound in which H is bonded to one or more electronegative nonmetals.

Binary compound – consists of two elements; it may be ionic or covalent.

Binding energy (nuclear binding energy) – the energy equivalent ($E = mc^2$) of the mass deficiency of an atom (where E is the energy in joules, m is the mass in kilograms, and c is the speed of light in m/s^2).

Boiling – the phase transition of a liquid vaporizing.

Boiling point – the temperature at which the vapor pressure of a liquid is equal to the applied pressure; the *condensation point*.

Boiling point elevation – the increase in the boiling point of a solvent caused by the dissolution of a nonvolatile solute.

Bomb calorimeter – a device used to measure the heat transfer between a system and its surroundings at constant volume.

Bond – the attraction and repulsion between atoms and molecules is a cornerstone of chemistry.

Bond energy – the amount of energy necessary to break one mole of bonds in a substance, dissociating the substance in its gaseous state into atoms of its elements in the gaseous state.

Bond order – half the number of electrons in bonding orbitals minus half the electrons in antibonding orbitals.

Bonding orbital – a molecular orbit lower in energy than any of the atomic orbitals from which it is derived; lends stability to a molecule or ion when populated with electrons.

Bonding pair – pair of electrons involved in a covalent bond.

Boron hydrides – binary compounds of boron and hydrogen.

Born-Haber cycle – a series of reactions (and the accompanying enthalpy changes) which, when summed, represent the hypothetical one-step reaction by which elements in their standard states are converted into crystals of ionic compounds (and the accompanying enthalpy changes).

Boyle's Law – at a constant temperature, the volume occupied by a definite mass of a gas is inversely proportional to the applied pressure.

Breeder reactor – a nuclear reactor that produces more fissionable nuclear fuel than it consumes.

Brønsted-Lowrey acid – a chemical species that donates a proton.

Brønsted-Lowrey base – a chemical species that accepts a proton.

Buffer solution – resists change in pH; contains either a weak acid and a soluble ionic salt of the acid or a weak base and a soluble ionic salt of the base.

Buret – a piece of volumetric glassware, usually graduated in 0.1 mL intervals, used to deliver solutions for titrations in a quantitative (drop-like) manner; also spelled *burette*.

C

Calorie – the amount of heat required to raise the temperature of one gram of water from 14.5 °C to 15.5 °C; 1 calorie = 4.184 joules.

Calorimeter – a device used to measure the heat transfer between a system and its surroundings.

Canal ray – a stream of positively charged particles (cations) that moves toward the negative electrode in cathode ray tubes; observed to pass through canals in the negative electrode.

Capillary – a tube having a very small inside diameter.

Capillary action – the drawing of a liquid up the inside of a small-bore tube when adhesive forces exceed cohesive forces; the depression of the surface of the liquid when cohesive forces exceed the adhesive forces.

Catalyst – a chemical compound used to change the rate (to speed or slow it) of a regenerated reaction (i.e., not consumed) at the end of the reaction.

Catenation – the bonding of atoms of the same element into chains or rings (i.e., the ability of an element to bond with itself).

Cathode – the electrode at which reduction occurs; in a cathode ray tube, the negative electrode.

Cathodic protection – protection of a metal (making a cathode) against corrosion by attaching it to a sacrificial anode of a more easily oxidized metal.

Cathode ray tube – a closed glass tube containing gas under low pressure, with electrodes near the ends and a luminescent screen near the positive electrode; produces cathode rays when a high voltage is applied.

Cation – a positive ion; an atom or group of atoms that has lost one or more electrons.

Cell potential – the potential difference, E_{cell}, between oxidation and reduction half-cells under nonstandard conditions; the force in a galvanic cell pulls electrons through a reducing agent to an oxidizing agent.

Central atom – an atom in a molecule or polyatomic ion bonded to more than one other atom.

Chain reaction – a reaction that, once initiated, sustains itself and expands; a reaction in which reactive species, such as radicals, are produced in more than one step, allowing these reactive species to propagate the chain reaction.

Charles' Law – at constant pressure, the volume occupied by a definite mass of gas is directly proportional to its absolute temperature.

Chemical bonds – the attractive forces holding atoms together in elements or compounds.

Chemical change – when one or more new substances are formed.

Chemical equation – description of a chemical reaction by placing the formulas of the reactants on the left of an arrow and the formulas of the products on the right.

Chemical equilibrium – a state of dynamic balance in which the rates of forward and reverse reactions are equal; there is no net change in concentrations of reactants or products while a system is at equilibrium.

Chemical kinetics – studies rates and mechanisms of chemical reactions and factors they depend on.

Chemical periodicity – the variations in properties of elements with their position in the periodic table.

Chemical reaction – the change of one or more substances into another or multiple substances.

Cloud chamber – a device for observing the paths of speeding particles as vapor molecules condense on them to form fog-like tracks.

Cobalt chloride paper – water test; water changes the color from blue to pink.

Coefficient of expansion – the ratio of the change in the length or the volume of a body to the original length or volume for a unit change in temperature.

Cohesive forces – the forces of attraction among particles of a liquid.

Colligative properties – physical properties of solutions that depend upon the number but not the kind of solute particles present.

Collision theory – the theory of reaction rates that states that effective collisions between reactant molecules must occur for the reaction to occur.

Colloid – a heterogeneous mixture in which solute-like particles do not settle (e.g., many kinds of milk).

Combination reaction – two substances (elements or compounds) combine to form one compound.

Combustible – classification of liquid substances that burn based on flashpoints; any liquid having a flashpoint at or above 37.8 °C (100 °F) but below 93.3 °C (200 °F), except any mixture having components with flashpoints of 93.3 °C (200 °F) or higher, the total of which makes up 99% or more of the volume of the mixture.

Combustion (or *burning*) – an exothermic reaction between an oxidant and fuel with heat and often light.

Common ion effect – suppression of ionization of a weak electrolyte by the presence in the same solution of a strong electrolyte containing one of the same ions as the weak electrolyte.

Complex ions – ions resulting from coordinating covalent bonds between simple ions and other ions or molecules.

Composition stoichiometry – describes the quantitative (mass) relationships among elements in compounds.

Compound – a substance of two or more chemically bonded elements in fixed proportions; can be decomposed into constituent elements.

Compressed gas – a single or mixture of gases having (in a container) an absolute pressure exceeding 40 psi at 21.1 °C (70 °F).

Compression – an area in a longitudinal wave where the particles are closer and pushed in.

Concentration – the amount of solute per unit volume, the mass of solvent or solution.

Condensation – the phase change from gas to liquid.

Condensed phases – the liquid and solid phases; phases in which particles interact strongly.

Condensed states – the solid and liquid states.

Conduction – heat transfer between substances in direct contact with each other (i.e., must be touching); when particles of a hotter substance vibrate, these molecules bump into nearby particles and transfer some energy.

Conduction band – a vacant or partially filled band of energy levels just higher in energy than a filled band; a band within which, or into which, electrons must be promoted to allow electrical conduction to occur in a solid.

Conductor – a material that allows electric flow more freely.

Conjugate acid-base pair – in Brønsted-Lowry terms, reactant and product that differ by a proton, H^+.

Conformations – structures of a compound that differ by the extent of rotation about a single bond.

Continuous spectrum – contains wavelengths in a specified region of the electromagnetic spectrum.

Control rods – rods of materials such as cadmium or boron steel that act as neutron absorbers (not merely moderators), used in nuclear reactors to control neutron fluxes and therefore fission rates.

Conjugated double bonds – double bonds separated from each other by one single bond $-C=C-C=C-$

Contact process – the industrial process for sulfur trioxide (SO_3) and sulfuric acid (H_2SO_4) production from sulfur dioxide (SO_2).

Convection – the physical flow of matter when heat flows by energized molecules from one place to another through the movement of fluids. Transfer of heat through a liquid or a gas occurs when molecules of the liquid or gas move and carry the heat.

Coordinate covalent bond – a covalent bond with shared electrons furnished by the same species; a bond between a Lewis acid and a Lewis base.

Coordination compound or complex – a compound containing coordinate covalent bonds.

Coordination number – the number of donor atoms coordinated to a metal; the number of nearest neighbors of an atom or ion in describing crystals.

Coordination sphere – the metal ion and its coordinating ligands, but no uncoordinated counter-ions.

Corrosion – oxidation of metals (e.g., rusting) in the presence of air and moisture.

Coulomb – the SI unit of electrical charge; unit symbol – C.

Covalent bond – a force of attraction (chemical bond) formed by the sharing of electron pairs between two atoms.

Covalent compounds – compounds made of two or more nonmetal atoms bonded by sharing valence electrons.

Critical mass – the minimum mass of a particular fissionable nuclide in a given volume required to sustain a nuclear chain reaction.

Critical point – the combination of critical temperature and critical pressure of a substance.

Critical pressure – the pressure required to liquefy a gas (vapor) at its *Critical temperature*.

Critical temperature – the temperature above which a gas cannot be liquefied; the temperature above which a substance cannot exhibit distinct gas and liquid phases.

Crystal – a solid packed with ions, molecules, or atoms in an orderly fashion.

Crystal field stabilization energy – a measure of the net energy of stabilization gained by a metal ion's nonbonding d electrons due to complex formation.

Crystal field theory – bonding in transition metal complexes in which ligands and metal ions are treated as point charges; a purely ionic model; ligand point charges represent the crystal (electrical) field perturbing the metal's d orbitals containing nonbonding electrons.

Crystal lattice – a pattern of arrangement of particles in a crystal.

Crystal lattice energy – the amount of energy that holds a crystal together; the energy change when a mole of solid forms from its constituent molecules or ions (for ionic compounds) in their gaseous state (consistently negative).

Crystalline solid – a solid characterized by a regular, ordered arrangement of particles.

Curie (Ci) – the basic unit to describe the intensity of radioactivity in a sample of material; one curie equals 37 billion disintegrations per second or approximately the amount of radioactivity given off by 1 gram of radium.

Current – a flow of charged particles, such as electrons or ions, moving through an electrical space or conductor. It is measured as the net rate of flow of electric charge; the unit is an Ampere (A).

Cuvette – glassware used in spectroscopic experiments; usually made of plastic, glass, or quartz, and should be as clean and transparent as possible.

Cyclotron – a device for accelerating charged particles along a spiral path.

D

Dalton's Law (or the *law of partial pressures*) – the pressure exerted by a mixture of gases is the sum of the partial pressures of the individual gases.

Daughter nuclide – nuclide produced in nuclear decay.

Debye – the unit used to express dipole moments.

Degenerate – in orbitals, describes orbitals of the same energy.

Deionization – the removal of ions; in the case of water, mineral ions such as sodium, iron, and calcium.

Deliquescence – substances that absorb water from the atmosphere to form liquid solutions.

Delocalization – in reference to electrons, bonding electrons distributed among more than two atoms bonded; occurs in species that exhibit resonance.

Density – mass per unit volume; $D = M \times V$.

Deposition – settling particles within a solution; the direct solidification of vapor by cooling; see *sublimation*.

Derivative – a compound that can be imagined arising from a parent compound by replacing one atom with another atom or group of atoms; used extensively in organic chemistry to identify compounds.

Detergent – a soap-like emulsifier with sulfate, SO_3, or a phosphate group instead of carboxylate group.

Deuterium – an isotope of hydrogen whose atoms are twice as massive as ordinary hydrogen; deuterium atoms contain a proton and a neutron in the nucleus.

Dextrorotatory – refers to an optically active substance that rotates plane-polarized light clockwise, also known as *dextro*.

Diagonal similarities – chemical similarities in the Periodic Table of Elements of Period 2 to elements of Period 3 one group to the right, especially evident toward the left of the periodic table.

Diamagnetism – weak repulsion by a magnetic field.

Differential Scanning Calorimetry (DSC) – a technique for measuring temperature, direction, and magnitude of thermal transitions in a sample material by heating/cooling and comparing the amount of energy required to maintain its rate of temperature increase or decrease with an inert reference material under similar conditions.

Differential Thermal Analysis (DTA) – a technique for observing the temperature, direction, and magnitude of thermally induced transitions in a material by heating/cooling a sample and comparing its temperature with an inert reference material under similar conditions.

Differential thermometer – a thermometer used to measure minimal temperature changes accurately.

Dilution – the process of reducing the concentration of a solute in a solution, usually by mixing with more solvent.

Dimer – a molecule formed by combining two smaller (identical) molecules.

Dipole – electric or magnetic separation of charge; charge separation between two covalently bonded atoms.

Dipole-dipole interactions – attractive electrostatic forces between polar molecules (i.e., between molecules with permanent dipoles).

Dipole moment – the product of the distance separating opposite charges of equal magnitude; a measure of the polarity of a bond or molecule; a measured dipole refers to the dipole moment of an entire molecule.

Dispersing medium – the solvent-like phase in a colloid.

Dispersed phase – the solute-like species in a colloid.

Displacement reactions – reactions in which one element displaces another from a compound.

Disproportionation reactions – redox reactions in which the oxidizing agent and the reducing agent are the same species.

Dissociation – in an aqueous solution, the process by which a solid ionic compound separates into its ions.

Dissociation constant – equilibrium constant for dissociating a complex ion into a simple ion and coordinating species (ligands).

Dissolution or solvation – the spread of ions in a monosaccharide.

Distilland – the material in a distillation apparatus that is to be distilled.

Distillate – the material in a distillation apparatus collected in the receiver.

Distillation – separating a liquid mixture into its components based on differences in boiling points; the process in which components of a mixture are separated by boiling away the more volatile liquid; the vaporization of a liquid by heating and then the condensation of the vapor by cooling.

Domain – a cluster of atoms in a ferromagnetic substance, which align in the same direction in the presence of an external magnetic field.

Donor atom – a ligand atom whose electrons are shared with a Lewis acid.

d-**orbitals** – beginning in the third energy level, a set of five degenerate orbitals per energy level, higher in energy than *s* and *p* orbitals of the same energy level.

Dosimeter – a small, calibrated electroscope worn by laboratory personnel to measure incident ionizing radiation or chemical exposure.

Double bond – covalent bond resulting from the sharing of four electrons (two pairs) between two atoms.

Double salt – a solid consisting of two co-crystallized salts.

Doublet – two peaks or bands of about equal intensity appearing close on a spectrogram.

Downs cell – an electrolytic cell for the commercial electrolysis of molten sodium chloride.

DP number – the degree of polymerization; the average number of monomer units per polymer unit.

Dry cells (voltaic cells) – ordinary batteries for appliances (e.g., flashlights, radios).

Dumas method – a method used to determine the molecular weights of volatile liquids.

Dynamic equilibrium – an equilibrium in which the processes occur continuously with no net change.

E

Earth metal – highly reactive elements in group IIA of the periodic table (includes beryllium, magnesium, calcium, strontium, barium, and radium); see *alkaline earth metal*.

Effective collisions – a collision between molecules resulting in a reaction, one in which the molecules collide with proper relative orientations and sufficient energy to react.

Effective molality – the sum of the molalities of solute particles in a solution.

Effective nuclear charge – the nuclear charge experienced by the outermost electrons of an atom; the actual nuclear charge minus the effects of shielding due to inner-shell electrons (e.g., a set of dx^2-y^2 and dz^2 orbitals); those d orbitals within a set with lobes directed along the x, y, and z axes.

Electrical conductivity – the measure of how easily an electric current can flow through a substance.

Electric charge – a measured property (coulombs) that determines electromagnetic interaction.

Electrochemical cell – using a chemical reaction, an electromotive force is generated.

Electrochemistry – the study of chemical changes produced by electrical current and electricity production by chemical reactions.

Electrodes – surfaces upon which oxidation and reduction half-reactions occur in electrochemical cells; a conductor dips into an electrolyte and allows the electrons to flow to and from the electrolyte.

Electrode potentials – potentials, E, of half-reactions as reductions vs. standard hydrogen electrode.

Electrolysis – 1) occurs in electrolytic cells; chemical decomposition occurs by passing an electric current through a solution containing ions. 2) producing a chemical change using electricity; used to split up water into H and O_2.

Electrolyte – a substance (i.e., anions, cations) which, when dissolved in water, conducts electricity. An ionic solution that conducts a certain amount of current is categorized into two types: weak and strong.

Electrolytic cells – electrochemical cells in which electrical energy causes nonspontaneous redox reactions to occur (i.e., forced to occur by applying an outside source of electrical energy).

Electrolytic conduction – electrical current passes through ions in a solution or pure liquid.

Electromagnetic radiation – energy propagated using electric and magnetic fields that oscillate in directions perpendicular to the direction of travel of the energy; a type of wave that can go through vacuums as well as material; classified as a "self-propagating wave."

Electromagnetism – fields of an electric charge and electric properties that change how particles move and interact.

Electromotive force – a device that gains energy as electric charges pass through it.

Electromotive series – the relative order of tendencies for elements and their simple ions to act as oxidizing or reducing agents; also known as the "activity series."

Electron – a subatomic particle having a mass of 0.00054858 amu and a charge of –1.

Electron affinity – the energy absorbed in the process in which an electron is added to a neutral isolated gaseous atom to form a gaseous ion with a 1– charge; it has a negative value if energy is released.

Electron configuration – the specific distribution of electrons in atomic orbitals of atoms or ions.

Electron-deficient compounds – at least one atom (other than H) shares fewer than eight electrons.

Electron shells – an orbital around an atom's nucleus with a fixed number of electrons (usually 2 or 8).

Electronic transition – the transfer of an electron from one energy level to another.

Electronegativity – a measure of the relative tendency of an atom to attract electrons to itself when chemically combined with another atom.

Electronic geometry – the geometric arrangement of orbitals containing the shared and unshared electron pairs surrounding the central atom or polyatomic ion.

Electrophile – positively charged or electron-deficient.

Electrophoresis – separating ions by migration rate and direction of migration in an electric field.

Electroplating – a metal is covered with another metal layer using electricity; plating a metal onto a (cathodic) surface by electrolysis.

Element – a substance that cannot be decomposed into simpler substances by chemical means; defined by its *atomic number*. A substance that cannot be split into simpler substances by chemical means.

Eluant (or eluent) – the solvent used in the process of elution, as in liquid chromatography.

Eluate – a solvent (or mobile phase) which passes through a chromatographic column and removes the sample components from the stationary phase.

Emission spectrum – the emission of electromagnetic radiation by atoms (or other species) resulting from electronic transitions from higher to lower energy states.

Empirical formula – the simplest whole-number ratio of atoms of each element present in a compound; also known as the *simplest formula*.

Emulsifying agent – a substance that coats the particles of the dispersed phase and prevents coagulation of colloidal particles; an emulsifier.

Emulsion – colloidal suspension of a liquid in a liquid.

Endothermic – describes processes that absorb heat energy (*H*).

Endothermicity – the absorption of heat by a system as the process occurs.

Endpoint – the point at which an indicator changes color and titration stops.

Energy – a system's ability to do work.

Enthalpy (*H*) – the heat content of a specific amount of substance; E= PV.

Entropy (*S*) – a thermodynamic state or property that measures the degree of disorder (i.e., randomness) of a system; the amount of energy not available for work in a closed thermodynamic system (usually denoted by S).

Enzyme – a protein that acts as a catalyst in biological systems.

Equation of state – an expression that describes the behavior of matter in a given state; the van der Waals equation describes the behavior of the gaseous state.

Equilibrium or chemical equilibrium – a state of dynamic balance with the rates of forward and reverse reactions equal; the state of a system where neither forward nor reverse reaction is thermodynamically favored.

Equilibrium constant – a quantity characterizing equilibrium position for a reversible reaction; its magnitude is equal to mass action expression at equilibrium; equilibrium, "K," varies with temperature.

Equivalence point – the point when chemically equivalent amounts of reactants have reacted.

Equivalent weight – an oxidizing or reducing agent whose mass gains (oxidizing agents) or loses (reducing agents) 6.022×10^{23} electrons in a redox reaction.

Evaporation – vaporization of a liquid below its boiling point.

Evaporation rate – the rate at which a particular substance vaporizes (evaporate) compared to the rate of a known substance, such as ethyl ether, especially useful for health and fire-hazard considerations.

Excited state – any state other than the ground state of an atom or molecule; see *ground state*.

Exothermic – describes processes that release heat energy (*H*).

Exothermicity – the release of heat by a system as a process occurs.

Explosive – a chemical or compound that causes a sudden, almost instantaneous release of pressure, gas, heat, and light when subjected to sudden shock, pressure, high temperature, or applied potential.

Explosive limits – the range of concentrations over which a flammable vapor mixed with the proper ratios of air will ignite or explode if a source of ignition is provided.

Extensive property – a property that depends upon the amount of material in a sample.

Extrapolate – to estimate the value of a result outside the range of a series of known values; a technique used in the standard additions calibration procedure.

F

Faraday constant – a unit of electrical charge widely used in electrochemistry and equal to ~ 96,500 coulombs; represents 1 mole of electrons, or the Avogadro number of electrons: 6.022×10^{23} electrons.

Faraday's law of electrolysis – a two-part law that Michael Faraday published about electrolysis. 1. the mass of a substance altered at an electrode during electrolysis is directly proportional to the quantity of electricity transferred at that electrode. 2. the mass of an elemental material altered at an electrode is directly proportional to the element's equivalent weight; one equivalent weight of a substance is produced at each electrode during the passage of 96,487 coulombs of charge through an electrolytic cell.

Fast neutron – a neutron ejected at high kinetic energy in a nuclear reaction.

Ferromagnetism – the ability of a substance to become permanently magnetized by exposure to an external magnetic field.

Flashpoint – the temperature at which a liquid yields enough flammable vapor to ignite; various recognized industrial testing methods exist; therefore, the method used must be specified.

Fluorescence – absorption of high-energy radiation and subsequent emission of visible light.

First Law of Thermodynamics – the amount of energy in the universe is constant (i.e., energy is neither created nor destroyed in ordinary chemical reactions and physical changes); known as the Law of Conservation of Energy.

Fluids – substances that flow freely; gases and liquids.

Flux – a substance added to react with the charge or a product of its reduction; in metallurgy, it is usually added to lower the melting point.

Foam – colloidal suspension of a gas in a liquid.

Formal charge – a method of counting electrons in a covalently bonded molecule or ion; it counts bonding electrons as though they were equally shared between the two atoms.

Formula – a combination of symbols that indicates the chemical composition of a substance.

Formula unit – the smallest repeating unit of a substance; the molecule for nonionic substances.

Formula weight – the mass of one formula unit of a substance in atomic mass units.

Fossil fuels – formed from the remains of plants and animals that lived millions of years ago.

Fractional distillation – when a fractioning column is used in a distillation apparatus to separate the components of a liquid mixture with different boiling points.

Fractional precipitation – removal of some ions from a solution by precipitation while leaving other ions with similar properties in the solution.

Free energy change – the indicator of the spontaneity of a process at constant T and P (e.g., if ΔG is negative, the process is spontaneous).

Free radical – a highly reactive chemical species carrying no charge and having a single unpaired electron in an orbital.

Freezing – phase transition from liquid to solid.

Freezing point depression – the decrease in the freezing point of a solvent caused by the presence of a solute.

Frequency – the number of repeating points on a wave that passes a given observation point per unit time; the unit is 1 hertz = 1 cycle per 1 second.

Fuel – any substance that burns in oxygen to produce heat.

Fuel cells – a voltaic cell that converts the chemical energy of a fuel and an oxidizing agent directly into electrical energy continuously.

G

Gamma (γ) ray – a highly penetrating type of nuclear radiation similar to X-ray radiation, except that it comes from within the nucleus of an atom and has higher energy; energy-wise, very similar to cosmic rays, except that cosmic rays originate from outer space.

Galvanic cell – a battery made from an electrochemical cell with two metals connected by a salt bridge.

Galvanizing – placing a thin layer of zinc on a ferrous material to protect the underlying surface from corrosion.

Gangue – sand, rock, and other impurities surrounding the mineral of interest in an ore.

Gas – a state of matter in which the particles have no definite shape or volume, though they fill their container.

Gay-Lussac's Law – the expression used for each of the two relationships named after the French chemist Joseph Louis Gay-Lussac concerning the properties of gases; more usually applied to his law of combining volumes.

Geiger counter – a gas-filled tube that discharges electrically when ionizing radiation passes through it.

Gel – colloidal suspension of a solid dispersed in a liquid; a semi-rigid solid.

Gibbs (free) energy – the thermodynamic state function of a system that indicates the amount of energy available for the system to do useful work at constant T and P; a value that indicates the spontaneity of a reaction (usually denoted by *G*).

Graham's Law – the rates of effusion of gases are inversely proportional to the square roots of their molecular weights or densities.

Ground state – the lowest energy state or most stable state of an atom, molecule, or ion; see *excited state*.

Group – a vertical column in the periodic table; known as a family.

H

Haber process – a process for the catalyzed industrial production of ammonia from N_2 and H_2 at high temperature and pressure.

Half-cell – the compartment in which the oxidation or reduction half-reaction occurs in a voltaic cell.

Half-life – the time required for half of a reactant to be converted into product(s); the time required for half of a given sample to undergo radioactive decay.

Half-reaction – the oxidation or the reduction part of a redox reaction.

Halogens – group VIIA elements: F, Cl, Br, I; halogens are nonmetals.

Hard water – water high in dissolved minerals that makes it difficult to form lather with soap.

Heat – a form of energy that flows between two samples of matter because of temperature differences.

Heat capacity – the amount of heat required to raise the temperature of a mass one degree Celsius.

Heat of condensation – the amount of heat that must be removed from one gram of vapor at its condensation point to condense the vapor with no change in temperature.

Heat of crystallization – the amount of heat that must be removed from one gram of a liquid at its freezing point to freeze it with no change in temperature.

Heat of fusion – the amount of heat required to melt one gram of a solid at its melting point with no change in temperature; usually expressed in J/g; the molar heat of fusion is the amount of heat required to melt one mole of a solid at its melting point with no change in temperature and is usually expressed in kJ/mol.

Heat of solution – the amount of heat absorbed in forming a solution that contains one mole of the solute; the value is positive if heat is absorbed (endothermic) and negative if heat is released (exothermic).

Heat of vaporization – the amount of heat required to vaporize one gram of a liquid at its boiling point with no change in temperature; usually expressed in J/g; the molar heat of vaporization is the amount of heat required to vaporize one mole of liquid at its boiling point with no change in temperature and is usually expressed as ion kJ/mol.

Heisenberg uncertainty principle – states that it is impossible to determine the momentum and position of an electron simultaneously with absolute accuracy.

Henry's Law – the gas pressure above a solution is proportional to concentration of the gas in solution.

Hess' Law of heat summation – the enthalpy change for a reaction is the same whether it occurs in one step or a series of steps.

Heterogeneous catalyst – exists in a different phase (solid, liquid, or gas) from the reactants; a contact catalyst.

Heterogeneous equilibria – equilibria involving species in more than one phase.

Heterogeneous mixture – a mixture that does not have uniform composition and properties throughout.

Heteronuclear – consisting of different elements.

High spin complex – crystal field designation for an outer orbital complex; t_{2g} and e_g orbitals are singly occupied before pairing occurs.

Homogeneous catalyst – in the same phase (solid, liquid, or gas) as the reactants.

Homogeneous equilibria – when *reagents* and *products* are in the same phase (gases, liquids, or solids).

Homogeneous mixture – a mixture which has uniform composition and properties throughout.

Homologous series – compounds with each member differing from the next by a specific number and kind of atoms.

Homonuclear – consisting of only one element.

Hund's rule – single electrons must occupy orbitals of a given sublevel before pairing begins; see *Aufbau* (or *building up*) *principle*.

Hybridization – mixing atomic orbitals to form a new set of atomic orbitals with the same electron capacity and properties and energies intermediate between the original unhybridized orbitals.

Hydrate – a solid compound that contains a definite percentage of bound water.

Hydrate isomers – crystalline complexes that differ in whether water exists inside or outside the coordination sphere.

Hydration – the reaction of a substance with water.

Hydration energy – the energy change accompanying the hydration of a mole of gas and ions.

Hydride – a binary compound of hydrogen.

Hydrocarbons – compounds that contain only carbon and hydrogen.

Hydrogen bond – a relatively strong dipole-dipole interaction (but still considerably weaker than the covalent or ionic bonds) between molecules containing hydrogen directly bonded to a small, highly electronegative atom, such as N, O or F.

Hydrogenation – the reaction in which hydrogen adds across a double or triple bond.

Hydrogen-oxygen fuel cell – hydrogen is the fuel (reducing agent), and oxygen is the oxidizing agent.

Hydrolysis – the reaction of a substance with water or its ions.

Hydrolysis constant – an equilibrium constant for a hydrolysis reaction.

Hydrometer – a device used to measure the densities of liquids and solutions.

Hydrophilic colloids – colloidal particles that repel water molecules.

I

Ideal gas – a hypothetical gas that obeys the postulates of the Kinetic Molecular Theory.

Ideal gas law – the product of pressure and the volume of an ideal gas is directly proportional to the number of moles of the gas and the absolute temperature. $PV = nRT$

Ideal solution – obeys Raoult's Law strictly.

Immiscible liquids – do not mix to form a solution (e.g., oil and water).

Indicators – for acid-base titrations, organic compounds that exhibit different colors in solutions of different acidities, determine the point at which the reaction between two solutes is complete.

Inert pair effect – characteristic of the post-transition minerals; the tendency of the electrons in the outermost atomic *s* orbital to remain un-ionized or unshared in compounds of post-transition metals.

Inhibitory catalyst – an inhibitor; a catalyst that decreases the rate of reaction.

Inner orbital complex – valence bond designation for a complex in which the metal ion utilizes d orbitals for one shell inside the outermost occupied shell in its hybridization.

Inorganic chemistry – a part of chemistry concerned with inorganic (non-carbon-based) compounds.

Insulator – a material that resists the flow of electric current or heat transfer; it does not allow heat to flow easily.

Insoluble compound – a substance that will not dissolve in a solvent, even after mixing.

Integrated rate equation – an expression giving the concentration of a reactant remaining after a specified time; it has a different mathematical form for different orders of reactants.

Intermolecular forces – forces between individual particles (atoms, molecules, ions) of a substance.

Ion – a molecule that has gained or lost electrons; an atom or a group of atoms carries an electric charge (Na^+).

Ion product for water – equilibrium constant for water ionization; $Kw = [H_3O^+]\cdot[^-OH] = 1.00 \times 10^{-14}$ at 25 °C.

Ionic bond – the electrostatic attraction between oppositely charged ions, resulting from a transfer of electrons.

Ionic bonding – chemical bonding resulting from transferring electrons from one atom or group.

Ionic compounds – compounds containing predominantly ionic bonding.

Ionic geometry – arrangement of atoms (not lone pairs of electrons) about the central atom of a polyatomic ion.

Ionization – the breaking up of a compound into separate ions; in an aqueous solution, the process by which a molecular compound reacts with water and forms ions.

Ionization constant – equilibrium constant for the ionization of a weak electrolyte.

Ionization energy – the minimum amount of energy required to remove the most loosely held electron of an isolated gaseous atom or ion.

Ion exchange – a method of removing hardness from water, it replaces the positive ions that cause the hardness with H^+ ions.

Ionization isomers – result from the interchange of ions inside and outside the coordination sphere.

Isoelectric – having the same electronic configurations.

Isomers – different substances with the same formula.

Isomorphous – refers to crystals having the same atomic arrangement.

Isotopes – two or more forms of atoms of the same element with different masses; atoms containing the same number of protons but different numbers of neutrons.

IUPAC – acronym for "International Union of Pure and Applied Chemistry."

J

Joule (J) – a unit of energy in the SI system; one joule is $1\ kg\cdot m^2/s^2$, which is 0.2390 calories.

K

K capture – absorption of a K shell (n = 1) electron by a proton as it is converted to a neutron.

Kelvin – a unit of measure for temperature based upon an absolute scale.

Kinetics – a subfield of chemistry specializing in reaction rates.

Kinetic energy (*KE*) – energy that matter processes through its motion.

Kinetic Molecular Theory – a theory that attempts to explain macroscopic observations on gases in microscopic or molecular terms.

　　　　

L

Lanthanides – elements 57 (lanthanum) through 71 (lutetium); grouped because of their similar behavior in chemical reactions.

Lanthanide contraction – a decrease in the radii of the elements following the lanthanides compared to what would be expected if there were no *f*-transition metals.

Latent heat – the energy absorbed or released when a substance changes state without changing temperature.

Lattice – a unique arrangement of atoms or molecules in a crystalline liquid or solid.

Law of combining volumes (Gay-Lussac's Law) – at constant temperature and pressure, the volumes of reacting gases (and any gaseous products) can be expressed as ratios of small whole numbers.

Law of conservation of energy – energy cannot be created or destroyed; it can only change form.

Law of conservation of matter – there is no detectable change in the quantity of matter during an ordinary chemical reaction.

Law of conservation of matter and energy – the amount of matter and energy in the universe is fixed.

Law of definite proportions (law of constant composition) – different samples of a pure compound contain the same elements in the same proportions by mass.

Law of partial pressures (or *Dalton's Law*) – the pressure exerted by a mixture of gases is the sum of the partial pressures of the individual gases.

Laws of thermodynamics – physical laws which define quantities of thermodynamic systems describe how they behave and (by extension) set certain limitations, such as perpetual motion.

Lead storage battery – a secondary voltaic cell used in most automobiles.

Leclanche cell – a common type of *dry cell*.

Le Chatelier's principle – states that a system at equilibrium, or striving to attain equilibrium, responds in such a way as to counteract any stress placed upon it; if stress (change of conditions) is applied to a system at equilibrium, the system will shift in the direction that reduces stress.

Leveling effect – acids stronger than the acid characteristic of the solvent react with the solvent to produce that acid; a similar statement applies to bases. The strongest acid (base) that can exist in a given solvent is the acid (base) characteristic of the solvent.

Levorotatory (or *levo*) – an optically active substance rotates plane-polarized light counterclockwise.

Lewis acid – any species that can accept a share in an electron pair.

Lewis base – any species that can make available a pair of electrons.

Lewis dot formula (electron dot formula) – representation of a molecule, ion, or formula unit by showing atomic symbols and only outer shell electrons.

Ligand – a Lewis base in a coordination compound.

Light – that portion of the electromagnetic spectrum visible to the naked eye; known as "visible light."

Limiting reactant – a substance that stoichiometrically limits the number of products formed.

Linear accelerator – a device used for accelerating charged particles along a straight line path.

Line spectrum – an atomic emission or absorption spectrum.

Linkage isomers – a particular ligand bonds to a metal ion through different donor atoms.

Liquid – a state of matter which takes the shape of its container.

Liquid aerosol – colloidal suspension of a liquid in a gas.

London dispersion forces – very weak and very short-range attractive forces between short-lived temporary (induced) dipoles; known as "dispersion forces."

Lone pair – a pair of electrons residing on one atom and not shared by other atoms; an unshared pair.

Low spin complex – crystal field designation for an inner orbital complex; contains electrons paired t_{2g} orbitals before e_g orbitals are occupied in octahedral complexes.

Lubricant – a substance capable of reducing friction (i.e., a force that opposes the direction of motion).

M

Magnetic field – a space around a magnet where magnetism can be detected.

Magnetic quantum number – quantum mechanical solution to a wave equation designating the orbital within a given set (s, p, d, f) in which an electron resides.

Manometer – a two-armed barometer.

Mass (m) – a measure of the amount of matter in an object; mass is usually measured in grams or kilograms.

Mass action expression – for a reversible reaction, aA + bB cC + dD; the product of the concentrations of the products (species on the right), each raised to the power corresponding to its coefficient in the balanced chemical equation, divided by the product of the concentrations of reactants (species on the left), each raised to the power corresponding to its coefficient in the balanced equation; at equilibrium the mass action expression is equal to K.

Mass deficiency – the amount of matter converted into energy when an atom forms from constituent particles.

Mass number – the sum of the numbers of protons and neutrons in an atom; an integer.

Mass spectrometer – an instrument that measures the charge-to-mass ratio of charged particles.

Matter – anything that has mass and occupies space.

Mechanism – the sequence of steps by which reactants are converted into products.

Melting point – the temperature at which liquid and solid coexist in equilibrium.

Meniscus – the shape assumed by the surface of a liquid in a cylindrical container.

Melting – the phase change from a solid to a liquid.

Metal – a chemical element that is a good conductor of electricity and heat and forms cations and ionic bonds with nonmetals; elements below and to the left of the stepwise division (metalloids) in the upper right corner of the periodic table; about 80% of known elements are metals.

Metallic bonding – bonding within metals due to the electrical attraction of positively charged metal ions for mobile electrons that belong to the crystal.

Metallic conduction – conduction of electrical current through a metal or along a metallic surface.

Metalloid – a substance with the properties of metals and nonmetals (B, Al, Si, Ge, As, Sb, Te, Po and At).

Metathesis reactions – reactions in which two compounds react to form two new compounds, with no changes in oxidation number; reactions in which the ions of two compounds exchange partners.

Method of initial rates – method of determining the rate-law expression by carrying out a reaction with different initial concentrations and analyzing the resultant changes in initial rates.

Methylene blue – a heterocyclic aromatic chemical compound with the molecular formula $C_{16}H_{18}N_3SCl$.

Miscible liquids – mix to form a solution (e.g., alcohol and water).

Miscibility – the ability of one liquid to mix with (dissolve in) another liquid.

Mixture – two or more different substances mingled together but not chemically combined. A sample of matter composed of two or more substances, each of which retains its identity and properties.

Moderator – a substance (e.g., deuterium, oxygen, paraffin) that slows fast neutrons upon collision.

Molality – a concentration expressed as the number of moles of solute per kilogram of solvent.

Molarity – the number of moles of solute per liter of solution.

Molar solubility – the number of moles of a solute that dissolves to produce a liter of a saturated solution.

Mole – a measurement of an amount of substance; a single mole contains approximately 6.022×10^{23} units or entities; abbreviated mol.

Molecule – a chemically bonded number of electrically neutral atoms.

Molecular equation – a chemical reaction in which formulas are written as if substances existed as molecules; only complete formulas are used.

Molecular formula – indicates the number of atoms present in a molecule of a molecular substance.

Molecular geometry – the arrangement of atoms (not lone pairs of electrons) around a central atom of a molecule or polyatomic ion.

Molecular orbital (*MO*) – resulting from the overlap and mixing of atomic orbitals on different atoms (i.e., a region where an electron can be found in a molecule, as opposed to an atom); an MO belongs to the molecule.

Molecular orbital theory – a theory of chemical bonding based upon postulated molecular orbitals.

Molecular weight – the mass of one molecule of a nonionic substance in atomic mass units.

Molecule – the smallest particle of a compound capable of stable, independent existence.

Mole fraction – the number of moles of a component in a mixture divided by the number of moles in the mixture.

Monoprotic acid – can form only one hydronium ion per molecule; may be strong or weak.

Mother nuclide – nuclide that undergoes nuclear decay.

N

Native state – refers to the occurrence of an element in an uncombined or free state in nature.

Natural radioactivity – spontaneous decomposition of an atom.

Neat – conditions with a liquid reagent or gas performed with no added solvent or co-solvent.

Nernst equation – corrects standard electrode potentials for nonstandard conditions.

Net ionic equation – results from canceling spectator ions and eliminating brackets from a total ionic equation.

Neutralization – the reaction of an acid with a base to form a salt and water; usually, the reaction of hydrogen ions with hydroxide ions to form water molecules.

Neutrino – a particle that travels at speeds close to the speed of light; created due to radioactive decay.

Neutron – a neutral unit or subatomic particle with no net charge and a mass of 1.0087 amu.

Nickel-cadmium cell (NiCd battery) – a dry cell where the anode is Cd, the cathode is NiO_2, and the electrolyte is basic.

Nitrogen cycle – the complex series of reactions by which nitrogen is slowly but continually recycled in the atmosphere, lithosphere, and hydrosphere.

Noble gases – elements of the periodic Group 0; He, Ne, Ar, Kr, Xe, Rn; known as "rare gases;" formerly called "inert gases."

Nodal plane – a region in which the probability of finding an electron is zero.

Nonbonding orbital – a molecular orbital derived only from an atomic orbital of one atom; it lends neither stability nor instability to a molecule or ion when populated with electrons.

Nonelectrolyte – a substance whose aqueous solutions do not conduct electricity.

Nonmetal – an element that is not metallic.

Nonpolar bond – a covalent bond in which electron density is symmetrically distributed.

Nuclear – of or about the atomic nucleus.

Nuclear binding energy – the energy equivalent of the mass deficiency; the energy released in forming an atom from the subatomic particles.

Nuclear fission – when a heavy nucleus splits into nuclei of intermediate masses, and protons are emitted.

Nuclear magnetic resonance spectroscopy – a technique that exploits the magnetic properties of specific nuclei; helps identify unknown compounds.

Nuclear reaction – involves a change in the composition of a nucleus and can emit or absorb a tremendous amount of energy.

Nuclear reactor – a system in which controlled nuclear fission reactions generate heat energy on a large scale, subsequently converted into electrical energy.

Nucleons – particles comprising the nucleus: protons and neutrons.

Nucleus – the tiny and dense, positively charged center of an atom containing protons and neutrons, as well as other subatomic particles; the net charge is positive.

Nuclides – refers to different atomic forms of elements; in contrast to isotopes, which refer only to different atomic forms of a single element.

Nuclide symbol – designation for an atom A/Z E, in which E is the symbol of an element, Z is its atomic number, and A is its mass number.

Number density – a measure of the concentration of countable objects (e.g., atoms, molecules) in space; the number per volume.

O

Octahedral – molecules and polyatomic ions with one atom in the center and six atoms at the corners of an octahedron.

Octane number – a number that indicates how smoothly a gasoline burns.

Octet rule – during bonding, atoms tend to reach an electron arrangement with eight electrons in the outermost shell. Many representative elements attain at least a share of eight electrons in their valence shells when they form molecular or ionic compounds; there are some limitations.

Open sextet – species with only six electrons in the highest energy level of the central element (many Lewis acids).

Orbital – may refer to an atomic orbital or a molecular orbital.

Organic chemistry – the chemistry of substances that contain carbon-hydrogen bonds.

Organic compound – substances that contain carbon.

Osmosis – when solvent molecules pass through a semi-permeable membrane from a dilute solution into a more concentrated solution.

Osmotic pressure – the hydrostatic pressure produced on the surface of a semi-permeable membrane.

Outer orbital complex – valence bond designation for a complex in which the metal ion utilizes d orbitals in the outermost (occupied) shell in hybridization.

Overlap – the interaction of orbitals on different atoms in the same region of space.

Oxidation – the addition of oxygen or the loss of electrons. An algebraic increase in the oxidation number may correspond to a loss of electrons.

Oxidation numbers – quantitative values used as mechanical aids in writing formulas and balancing equations; for single-atom ions, they correspond to the charge on the ion; more electronegative atoms are assigned negative oxidation numbers, known as *oxidation states*.

Oxidation-reduction reactions – reactions in which oxidation and reduction occur; known as *redox reactions*.

Oxide – a binary compound of oxygen.

Oxidizing agent – the substance that oxidizes another substance and is reduced.

P

Pairing – a favorable interaction of two electrons with opposite m values in the same orbital.

Pairing energy – the energy required to pair two electrons in the same orbital.

Paramagnetism – attraction toward a magnetic field, stronger than diamagnetism but still weak compared to ferromagnetism.

Partial pressure – the force exerted by one gas in a mixture of gases.

Particulate matter – fine, divided solid particles suspended in polluted air.

Pauli exclusion principle – no electrons in the same atom may have identical sets of four quantum numbers.

Percentage ionization – the percentage of the weak electrolyte that will ionize in a solution of given concentration.

Percent by mass – 100% times the actual yield divided by the theoretical yield.

Percent composition – the mass percent of each element in a compound.

Percent purity – the percent of a specified compound or element in an impure sample.

Period – the elements in a horizontal row of the periodic table.

Periodicity – regular periodic variations of properties of elements with their atomic number (and position in the periodic table).

Periodic Law – the properties of the elements are periodic functions of their atomic numbers.

Periodic table – an arrangement of elements by increasing atomic numbers, emphasizing periodicity.

Peroxide – a compound with oxygen in –1 oxidation state; metal peroxides contain the peroxide ion, O_2^{2-}.

pH – the measure of acidity (or basicity) of a solution; negative logarithm of the concentration (mol/L) of the H_3O^+ [H^+] ion; scale is commonly used over a range 0 to 14.

Phase diagram – shows the equilibrium temperature-pressure relationships for different phases of a substance.

pH scale – a range from 0 to 14. If the pH of a solution is 7, it is neutral; if the pH of a solution is less than 7, it is acidic; if the pH of a solution is greater than 7, it is basic.

Permanent hardness – hardness (relative to lathering soap) in water that cannot be removed by boiling; caused by calcium sulfate.

Photoelectric effect – emission of an electron from the surface of a metal caused by impinging electromagnetic radiation of specific minimum energy; the current increases with increasing radiation intensity.

Photon – a carrier of electromagnetic radiation of all wavelengths, such as gamma rays and radio waves; known as a *quantum of light*.

Physical change – when a substance changes from one physical state to another, but no substances with different compositions are formed; physical change may involve a phase change (e.g., melting, freezing) or another physical change, such as crushing a crystal or separating one volume of liquid into different containers; it does not produce a new substance.

Plasma – a physical state of matter that exists at extremely high temperatures in which molecules are dissociated, and most atoms are ionized.

Polar bond – a covalent bond with an unsymmetrical distribution of electron density.

Polarimeter – a device used to measure optical activity.

Polarization – the buildup of a product of oxidation or reduction of an electrode, preventing further reaction.

Polydentate – refers to ligands with more than one donor atom.

Polyene – a compound that contains more than one double bond per molecule.

Polymerization – the combination of many small molecules to form large molecules.

Polymer – a large molecule consisting of chains or rings of linked monomer units, usually characterized by high melting and boiling points.

Polymorphous – refers to substances that can crystallize in more than one crystalline arrangement.

Polyprotic acid – forms two or more H_3O+ ions per molecule; often, at least one ionization step is weak.

Positron – a nuclear particle with the mass of an electron but opposite charge (positive).

Potential difference (or *voltage*) – the force that moves the electrons around circuit; unit is Volt (V).

Potential energy (*PE*) – energy stored in a body or a system due to its position in a force field or configuration.

Power – the rate at which energy is converted from one form to another; the unit is Watts (W). Power = voltage × current (P = VI).

Precipitate – an insoluble solid formed by mixing in solution the constituent ions of a slightly soluble solution.

Precision – how close the results of multiple experimental trials are; see *accuracy*.

Pressure – force per unit area; unit is Pascal (Pa).

Primary standard – a known high degree of purity substance that undergoes one invariable reaction with the other reactant of interest.

Primary voltaic cells – voltaic cells that cannot be recharged; no further chemical reaction is possible once the reactants are consumed.

Products – chemicals produced (from reactants) in a chemical reaction.

Proton – a subatomic particle having a mass of 1.0073 amu and a charge of +1, found atom's nucleus.

Protonation – the addition of a proton (H^+) to an atom, molecule, or ion.

Pseudobinaryionic compounds – contain more than two elements but are named like binary compounds.

Q

Quanta – the minimum amount of energy emitted by radiation.

Quantum mechanics – the study of how atoms, molecules, subatomic particles, etc., behave and are structured; a mathematical method of treating particles based on quantum theory, which assumes that energy (of small particles) is not infinitely divisible.

Quantum numbers – numbers that describe the energies of electrons in atoms; derived from quantum mechanical treatment.

Quarks – elementary particles and a fundamental constituent of matter, combining to form hadrons (i.e., protons and neutrons).

R

Radiation – 1) heat transfer through invisible rays, which travel outwards from the hot object without a medium. 2) high-energy particles or rays emitted during the nuclear decay processes.

Radical – an atom or group of atoms that contains one or more unpaired electrons; usually a very reactive species.

Radioactive dating – a method of dating ancient objects by determining the ratio of mother and daughter nuclides present in an object and relating the ratio to the object's age via half-life calculations.

Radioactive tracer – a small amount of radioisotope replacing a nonradioactive isotope of the element in a compound whose path (e.g., in the body) or whose decomposition products are monitored by detection of radioactivity; known as a "radioactive label."

Radioactivity – the spontaneous disintegration of atomic nuclei.

Raoult's Law – the vapor pressure of a solvent in an ideal solution decreases as its mole fraction decreases.

Rate-determining step – the slowest step in a mechanism; determines the overall reaction rate.

Rate-law expression – an equation relating the reaction rate to the concentrations of the reactants and the specific rate of the reaction.

Rate of reaction – the change in the concentration of a reactant or product per unit time.

Reactants – substances consumed in a chemical reaction; react together in a chemical reaction.

Reaction quotient – the mass action expression under any set of conditions (not necessarily equilibrium); its magnitude relative to K determines the direction in which the reaction must occur to establish equilibrium.

Reaction ratio – the relative amounts of reactants and products involved in a reaction; may be the ratio of moles, millimoles, or masses.

Reaction stoichiometry – describes the quantitative relationships among substances participating in chemical reactions.

Reactivity series (or activity series) – an empirical, calculated, and structurally analytical progression of a series of metals, arranged by "reactivity" from highest to lowest; used to summarize information about the reactions of metals with acids and water, double displacement reactions, and the extraction of metals from ores.

Reagent – a substance (or compound) added to a system to cause a chemical reaction or to visualize if a reaction occurs; the terms reactant and reagent are often used interchangeably; however, a *reactant* is more specifically a substance consumed during a chemical reaction.

Reducing agent – a substance that reduces another substance and is itself oxidized.

Reduction – the removal of oxygen or the gaining of electrons.

Resonance – the concept in which two or more equivalent dot formulas for the same arrangement of atoms (resonance structures) are necessary to describe the bonding in a molecule or ion.

Reverse osmosis – forcing solvent molecules to flow through a semi-permeable membrane from a concentrated solution into a dilute solution by applying greater hydrostatic pressure on the concentrated side than the osmotic pressure opposing it.

Reversible reaction – processes that do not go to completion and occur in the forward and reverse direction.

S

Saline solution – a general term for NaCl (i.e., sodium chloride) in water.

Salt – when a metal replaces the hydrogen of an acid.

Salts – ionic compounds composed of anions and cations.

Salt bridge – a U-shaped tube containing an electrolyte, connects the two half-cells of a voltaic cell.

Saturated solution –no more solute will dissolve at that temperature.

s-block elements – group 1 and 2 elements (alkali and alkaline metals), including hydrogen and helium.

Schrödinger equation – quantum state equation representing the behavior of an electron around an atom; describes the wave function of a physical system evolving.

Second Law of Thermodynamics – the universe tends toward a state of greater disorder in spontaneous processes.

Secondary standard – a solution that has been titrated against a primary standard; a standard solution.

Secondary voltaic cells – voltaic cells that can be recharged; original reactants can be regenerated by reversing the direction of the current flow.

Semiconductor – a substance that does not conduct electricity at low temperatures but will do so at higher temperatures.

Semi-permeable membrane – a thin partition between two solutions through which specific molecules can pass but others cannot.

Shielding effect – electrons in filled sets of s, p orbitals between the nucleus and outer shell electrons shield the outer shell electrons somewhat from the effect of protons in the nucleus; known as the "screening effect."

Sigma (σ) bonds – bonds resulting from the head-on overlap of atomic orbitals. The region of electron sharing is along and (cylindrically) symmetrical to the imaginary line connecting the bonded atoms.

Sigma orbital – molecular orbital resulting from the head-on overlap of two atomic orbitals.

Single bond – covalent bond resulting from the sharing of two electrons (one pair) between two atoms.

Sol – a suspension of solid particles in a liquid; artificial examples include sol-gels.

Solid – one of the states of matter, where the molecules are packed closely, resistance to movement/deformation, and volume change.

Solubility product constant – equilibrium constant for the dissolution of a slightly soluble compound.

Solubility product principle – the solubility product constant expression for a slightly soluble compound is the product of the concentrations of the constituent ions; each raised to the power that corresponds to the number of ions in one formula unit.

Solute – the dispersed (i.e., dissolved) phase of a solution; the solution is mixed into the solvent (e.g., $NaCl$ in saline water).

Solution – a homogeneous mixture of multiple substances; comprised of solutes and solvents; a mixture of a solute (usually a solid) and a solvent (usually a liquid).

Solvation – the process by which solvent molecules surround and interact with solute ions or molecules.

Solvent – the dispersing medium of a solution (e.g., H_2O in saline water).

Solvolysis – the reaction of a substance with the solvent in which it is dissolved.

s-orbital – a spherically symmetrical atomic orbital; one per energy level.

Specific gravity – the ratio of the density of a substance to the density of water.

Specific heat – the amount of heat required to raise the temperature of one gram of substance 1 °C.

Specific rate constant – an experimentally determined (proportionality) constant; different for different reactions, and which changes only with temperature; k in the rate-law expression: Rate = k [A] $\times$ [B].

Spectator ions – ions in a solution that do not participate in a chemical reaction.

Spectral line – any of several lines corresponding to definite wavelengths of an atomic emission or absorption spectrum; marks the energy difference between two energy levels.

Spectrochemical series – arrangement of ligands in order of increasing ligand field strength.

Spectroscopy – the study of radiation and matter, such as X-ray absorption and emission spectroscopy.

Spectrum – display of component wavelengths (colors) of electromagnetic radiation.

Speed of light – the speed at which radiation travels through a vacuum (299,792,458 m/sec).

Square planar – describes molecules and polyatomic ions with one atom in the center and four atoms at the corners of a square.

Square planar complex – relationship with metal in the center of a square plane, with ligand donor atoms at each of the four corners.

Standard conditions for temperature and pressure (STP) – used to compare experimental results (25 °C and 100.000 kPa).

Standard electrodes – half-cells in which the oxidized and reduced forms of a species are present at the unit activity (1.0 M solutions of dissolved ions, 1.0 atm partial pressure of gases, pure solids, and liquids).

Standard electrode potential – by convention, the potential (Eo) of a half-reaction as a reduction relative to the standard hydrogen electrode when species are present at unit activity.

Standard entropy – the absolute entropy of a substance in its standard state at 298 K.

Standard molar enthalpy of formation – the amount of heat absorbed in forming one mole of a substance in a specified state from its elements in their standard states.

Standard molar volume – space occupied by 1 mole of ideal gas under standard conditions; 22.4 liters.

Standard reaction – a process where the numbers of moles of reactants in the balanced equation, in their standard states, are entirely converted to the numbers of moles of products in the balanced equation, at their standard state.

State of matter – a homogeneous, macroscopic phase (e.g., gas, plasma, liquid, solid) in increasing concentration.

Stoichiometry – quantitative relationships of elements and compounds undergoing chemical changes.

Strong electrolyte – a substance that conducts electricity well in a dilute aqueous solution.

Strong field ligand – a ligand that exerts a strong crystal or ligand electrical field and generally forms low-spin complexes with metal ions when possible.

Structural isomers – compounds that contain the same number and kinds of atoms, but with different geometries.

Subatomic particles – comprise an atom (e.g., protons, neutrons, electrons).

Sublimation – the direct vaporization of a solid by heating without passing through the liquid state; a phase transition from solid to gas.

Substance – any matter, specimens with the same chemical composition and physical properties.

Substitution reaction – a reaction in which another atom or group of atoms replaces an atom or a group of atoms.

Supercooled liquids – liquids that, when cooled, apparently solidify but continue to flow very slowly under the influence of gravity.

Supercritical fluid – a substance at a temperature above its critical temperature.

Supersaturated solution – contains a higher than saturation concentration of solute; slight disturbance or seeding causes crystallization of excess solute.

Suspension – a heterogeneous mixture in which solute-like particles settle out of the solvent-like phase sometime after their introduction. A mixture of a liquid and a finely divided insoluble solid.

T

Talc – a mineral representing the Mohs Scale, composed of hydrated magnesium silicate with the chemical formula $H_2Mg_3(SiO_3)_4$ or $Mg_3Si_4O_{10}(OH)_2$.

Temperature – a measure of heat intensity (i.e., hotness or coldness of a sample); a measure of kinetic energy of an object. Units are measured in degrees, and scales include Celsius, Fahrenheit, and Kelvin.

Temporary hardness – hardness in water, removed by boiling; caused by calcium hydrogen carbonate.

Ternary acid – a ternary compound containing H, O and another element, often a nonmetal.

Ternary compound – a compound consisting of three elements; may be ionic or covalent.

Tetrahedral – a term used to describe molecules and polyatomic ions with one atom in the center and four atoms at the corners of a tetrahedron.

Theoretical yield – the maximum amount of a specified product that could be obtained from specified amounts of reactants, assuming complete consumption of the limiting reactant according to only one reaction and complete recovery of the product; see *actual yield*.

Theory – a model describing the nature of a phenomenon.

Thermal conductivity – a property of a material to conduct heat (often noted as k).

Thermal cracking – decomposition by heating a substance in the presence of a catalyst and without air.

Thermochemistry – the study of absorption/release of heat within a chemical reaction; studies heat energy associated with chemical reactions and physical transformations.

Thermodynamics – studying the effects of changing temperature, volume, or pressure (or work, heat, and energy) on a macroscopic scale.

Thermodynamic stability – when a system is in its lowest energy state with its environment (equilibrium).

Thermometer – a device that measures the average energy of a system.

Thermonuclear energy – energy from nuclear fusion reactions.

Third Law of Thermodynamics – entropy of a pure crystalline substance at absolute zero equals zero.

Titration – a procedure in which one solution is added to another solution until the chemical reaction between the two solutions is complete; the concentration of one solution is known, and that of the other is unknown. The process of adding one solution to a measured amount of another to find out exactly how much of each is required to react.

Torr – a unit to measure pressure; 1 Torr is equivalent to 133.322 Pa or 1.3158×10^{-3} atm.

Total ionic equation – the expression for a chemical reaction written to show the predominant form of species in aqueous solution or contact with water.

Transition elements (metals) – B Group elements except IIB in the periodic table; sometimes called transition elements, elements with incomplete d sub-shells; the d-block elements.

Transition state theory –reactants pass through high-energy transition states before forming products.

Transuranic element – an atomic number greater than 92; none of the transuranic elements are stable.

Triple bond – the sharing of three pairs of electrons within a covalent bond (e.g., N_2).

Triple point – where the temperature and pressure of three phases are the same; water has a unique phase diagram.

Tyndall effect – results from light scattering by colloidal particles (a mixture where one substance is dispersed evenly throughout another) or by suspended particles.

U

Uncertainty –any measurement that involves estimating any amount that cannot be precisely reproduced.

Uncertainty principle – knowing the location of a particle makes the momentum uncertain, while knowing the momentum of a particle makes the location uncertain.

Unit cell – the smallest repeating unit of a lattice.

Unit factor – statements used in converting between units.

Universal (or ideal) gas constant – proportionality constant in the ideal gas law (0.08206 L·atm/(K·mol)).

UN number – a four-digit code used to note hazardous and flammable substances.

Unsaturated hydrocarbons – hydrocarbons that contain double or triple carbon-carbon bonds.

V

Valence bond theory – proposes that covalent bonds are formed when atomic orbitals on different atoms overlap and the electrons are shared.

Valence electrons – outermost electrons of atoms; usually those involved in bonding.

Valence shell electron pair repulsion theory (VSEPR) – assumes electron pairs are arranged around the central element of a molecule or polyatomic ion with maximum separation (and minimum repulsion) among regions of high electron density.

Valency – the number of electrons an atom wants to gain, lose, or share to have a full outer shell.

Van der Waals' equation – a quantitative relationship of a state extending the ideal gas law to real gases by including two empirically determined parameters, which are specific for different gases.

Van der Waals force – one of the forces (attraction/repulsion) between molecules.

Van't Hoff factor – the ratio of moles of particles in solution to moles of solute dissolved.

Vapor – when a substance is below the critical temperature in the gas phase.

Vaporization – the phase change from liquid to gas.

Vapor pressure – the particle pressure of vapor at the surface of its parent liquid.

Viscosity – the resistance of a liquid to flow (e.g., oil has a higher viscosity than water).

Volt – one joule of work per coulomb; the unit of electrical potential transferred.

Voltage – the potential difference between two electrodes; a measure of the chemical potential for a redox reaction.

Voltaic cells – electrochemical cells in which spontaneous chemical reactions produce electricity; known as *galvanic cells*.

Voltmeter – an instrument that measures the cell potential.

Volumetric analysis – measuring the volume of a solution (of known concentration) to determine the substance's concentration within the solution; see *titration*.

W

Water equivalent – the amount of water absorbing the same heat as the calorimeter per degree of temperature increase.

Weak electrolyte – a substance that conducts electricity poorly in a dilute aqueous solution.

Weak field ligand – a ligand that exerts a weak crystal or ligand field and generally forms high-spin complexes with metals.

X

X-ray – electromagnetic radiation between gamma and UV rays.

X-ray diffraction – a method for establishing structures of crystalline solids using single-wavelength X-rays and studying the diffraction pattern.

X-ray photoelectron spectroscopy – a spectroscopic technique used to measure the composition of a material.

Y

Yield – the amount of product produced during a chemical reaction.

Z

Zone melting – remove impurities from an element by melting and slowly traveling it down an ingot (cast).

Zone refining – a method of purifying a metal bar by passing it through an induction heater; this causes impurities to move along a melted portion.

Zwitterion (formerly called a dipolar ion) – a neutral molecule with a positive and negative electrical charge; multiple positive and negative charges can be present, distinct from dipoles at different locations within that molecule; known as *inner salts.*

Frank J. Addivinola, Ph.D.

This study guide's lead author and chief editor is Dr. Frank Addivinola. With his outstanding education, laboratory research, and decades of university science teaching, Dr. Addivinola lent his expertise to oversee the development of this series.

During his extensive career, Dr. Addivinola held faculty positions at colleges and universities, including Harvard University, Johns Hopkins University, University of Maryland, and Northeastern University, and taught undergraduate and graduate-level courses in biology, biochemistry, organic chemistry, inorganic chemistry, physics, anatomy and physiology, medical terminology, nutrition, and medical ethics. He received several awards for his research and presentations.

Dr. Frank Addivinola conducted original research in developmental biology as a doctoral candidate and pre-IRTA fellow in Molecular and Cell Biology at the National Institutes of Health (NIH). His dissertation advisor was Nobel laureate Marshall W. Nirenberg, Chief of the Biochemical Genetics Laboratory at the National Heart, Lung, and Blood Institute (NHLBI). Before NIH, Dr. Addivinola researched prostate cancer in the Cell Growth and Regulation Laboratory of Dr. Arthur Pardee at the Dana Farber Cancer Institute of Harvard Medical School.

Dr. Addivinola holds an undergraduate degree in biology from Williams College. He completed his Masters at Harvard University, Masters in Biotechnology at Johns Hopkins University, and five other graduate degrees at the University of Maryland University College, Suffolk University, and Northeastern University.

Electronic Structure & Periodic Table

Chemical Bonding

States of Matter & Phase Equilibria

Stoichiometry

Solution Chemistry

Chemical Kinetics & Equilibrium

Acids & Bases

Chemical Thermodynamics

Electrochemistry

Visit our Amazon store

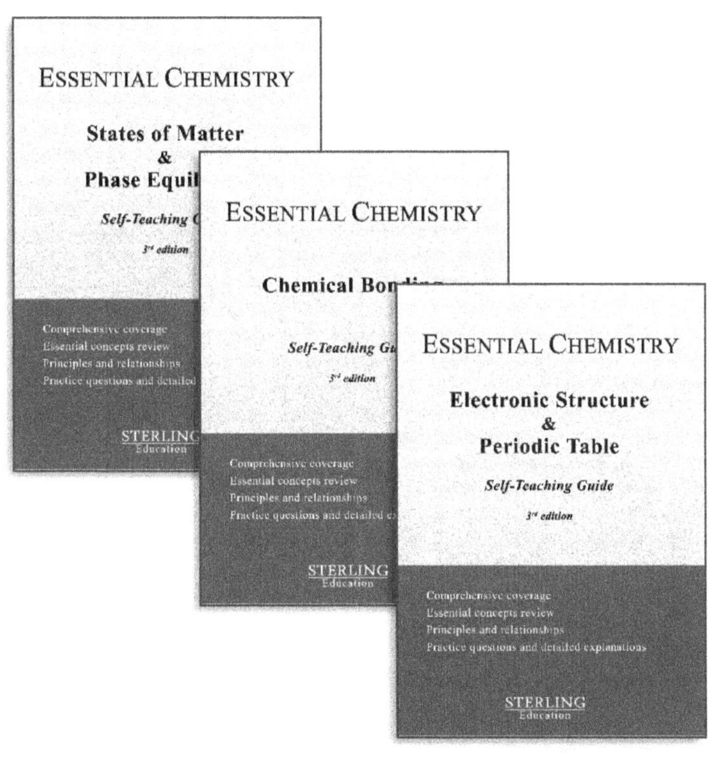

For online practice resources visit
https://www.sterling-prep.com

If you benefited from this book, please leave a review on Amazon so others
can learn from your input. Reviews help us understand our customers'
needs and experiences while keeping our commitment to quality.